Towards Sustainable New Towns?

Interconnected Experiences Spanning the North and South Mediterranean

P.I.E. Peter Lang

Bruxelles · Bern · Berlin · Frankfurt am Main · New York · Oxford · Wien

Pascaline GABORIT (ed.)

Towards Sustainable New Towns?

Interconnected Experiences Spanning the North and South Mediterranean

This publication has been co-financed by the European Commission

We would likc to thank:

The CIUDAD programme of the European Union, the CA Marne la Vallée-Val Maubuée, leader of New Medina project, the partners of the New Medina project: Touiza Solidarity, the ENTP, the EPA VNSA, Al Omrane Ch'Rafate and the HBRC, the experts who participated in the various panels.

The authors of the book.

For reviewing the content and the layout: Edoardo Guglielmetti, Stella Karavazaki, Susanne Wander and Phoebe Stirling.

For the translation: Vincent Vermeylen, David Bywell, Diana Molina and Jonathan Bitoun.

The opinions expressed in this publication are those of the authors and do not necessarily reflect the standpoint of the programme.

Cover Picture: New Fayoum, © ENTP, 2012.

Éditions scientifiques internationales
Brussels, 2013
1 avenue Maurice, B-1050 Brussels, Belgium
www.peterlang.com; info@peterlang.com

Printed in Germany

ISBN 978-2-87574-061-8
D/2013/5678/32

CIP available from the Library of Congress, USA, and the British Library, GB

Bibliographic information published by "Die Deutsche Nationalbibliothek"

"Die Deutsche Nationalbibliothek" lists this publication in the "Deutsche Nationalbibliografie"; detailed bibliographic data is available on the Internet at <http://dnb.ddb.de>.

Table of Contents

Foreword

One of the fundamental objectives of this work is to interrogate the notion of the new city: to ask "why" and "for whom" the new city exists. To adopt a term commonly used today, this is a question of the governance of such a project.

The design and construction of a functioning city cannot be left to the sole care of development and buildings experts, though without them nothing is possible. Making something of a city is not simply a case of technical expertise and should not be left to the "knowledgeable" alone. Cities are not built for the sake of argument but rather out of necessity, in response to the hopes of human beings who will live there, of companies planning to start business there, of elected representatives looking to manage its institutions and processes.

Governing such operations is therefore vital and must as rapidly as possible bring together, with equal rights and duties, all people concerned. To do this there is a need for all those involved – development experts, elected representatives, existing residents, if there are any, gradually the new residents... – to have, from the very outset of the operation, a clear vision of what's at stake. It is absolutely key that information be distributed as widely as possible and that all those involved speak the same language. In most cases, those lacking in knowledge about the processes at work need to be brought up to speed: training is therefore essential for many people involved in complex projects such as building a new city.

Finally, while participation from everyone is paramount, the fact remains that this type of involved development requires a strong political will, whether at central, regional or local government level. Following the discussion phase of looking at feasible choices comes the time for decision, if we want to avoid getting bogged down and facing inevitable financial battles.

The issue revolves around North-South exchanges on these questions between local governments and experts.

The New Medina project in particular, as the source of this work, has resulted in a fruitful encounter between players from Europe's new cities and women and men from the southern Mediterranean who have embarked today on a process to build new urban entities. The Algerian,

Moroccan and Egyptian people involved are highly trained and leading experts in all development techniques and processes.

However, being at the start of their development operation, they can learn from our hesitations and past mistakes, perhaps enabling them to avoid repetition and to save time and gain confidence in the task they complete. Even more than the Europe/Southern Mediterranean dialogue, the South/South dialogue is what matters.

During the panel sessions in the New Medina project, the Algerian, Moroccan and Egyptian professionals and experts listened to and understood each other on many points, even though diverging viewpoints did remain. This is fundamental for progress in urban projects in the Southern Mediterranean, since, even though a universal urban model does not and will never exist, the climate, geological, institutional, even political conditions unite, beyond any existing differences, stakeholders in the South.

It is not necessarily up to a European professional to say how to construct in desert areas!

Likewise, faced with similar situations and concerns as well as common difficulties, Moroccan developers have decided to get together and share their knowledge and experiences with each other, so that each of them can go on to produce better work in their respective places.

Looking beyond the work completed during the New Medina project, naturally marked in time, it would appear vital for these collaborations and exchanges to be able to continue, given that development itself is long-term.

We know only too well that what is created at the birth of a new city will not be what the city becomes ten or twenty years down the track. The construction of a new urban entity takes decades, during which time economic, social and political environments evolve constantly. We therefore need to understand that the base model is only a model used to establish a framework that will evolve constantly, otherwise, the only certainty will be a failed operation. This work addresses all these key themes and proposes several pragmatic avenues and thoughts to explore. It would clearly be greatly beneficial to be able to pursue and delve further into all the work already carried out: why not under a new project?

Jean-Pierre MARCHETTI

Senior Expert in International Urban Development

Former Managing Director of the New Agglomeration
Association for Marne-la-Vallée/Val Maubuée

Introduction

When space leads to continuity and memory reflects places, the city provides bearings.
Thought[1]

Why write and speak on a topic as specific, technical, problematic, debated and difficult as new cities? Understood as cities created from scratch and *ex nihilo* based on a development plan, they serve to break away from historic cities that experience more organic, spontaneous and straightforward developments and transformations as well as complex and multifaceted growth that could well be compared to a slice of multi-layered "pastry". New cities represent special cases in terms of urban development, with structured master plan, development and organisation projects for space. They are perilous adventures, striving to secure a special status at the suburban and peri-urban crossroads.

These recent or brand new cities are also a reflection of different contexts and trends. They bring together a whole host of political, structural and technical issues in one place and call on a vast array of expertise all busily working together to take up an enormous challenge: creating cities from start to finish. Manufacturing these cities means developing space, including social areas, the environment and places for people to live together. "*Space, for example a city neighbourhood, governs how its residents are grouped together, how they spread out, squeeze together or how they disperse and move around... As such, the fact of belonging to the city or to the countryside does not trigger the same consciousness among individuals of belonging to a group, due to changes in how social activities are set up*".[2]

In new cities, the lack of historical roots prevents, on the surface at least, the ability to lay claim to a collective and spatial memory. Sudden changes to the physical environment put a stop to the idea of continuity as conveyed by older cities, but do they in fact prevent a sense of belonging? Since the memory of newly created space is closely tied into

1 Original: *Quand l'espace renvoie à la continuité et la mémoire aux lieux, c'est que la ville donne des repères*. Thought from the authors.

2 Marcel J.C. in Deloye Y. and Haroche C., *Maurice Halbwachs: Espaces Mémoire et psychologie collective*, Paris, publication de la Sorbonne, 2004, pp. 53-54.

the experiences of its first residents, its intensity could indeed serve as a substitute for the lack of years of existence.

The rapid development of new cities does however bring with it several risks in terms of sameness, absence of contrast, difficulties putting services in place for the benefit of people, in sum, creating real social spaces.

The different issues linked to creating more cities have existed and sometimes continue to exist on both sides of the Mediterranean. At the heart of the challenge, exchanges between cities and experts add further value to discussions and local actions.

For countries on the southern side of the Mediterranean, the creation of new cities has become one of the priorities of master plans and urban policies. In Morocco and Algeria, and more widely in Egypt, they have sprouted up on the outskirts of large cities or in the desert, in an attempt to address the growing need for housing, to the detriment of locally-inspired projects, originating in the community.

"New Medina" is a European project funded by the CIUDAD[3] programme. The project has two goals: the first is to enable cities in the North and South to reflect on the pioneering experiences of new cities, identifying good practices and also mistakes notably in terms of incorporating environmental principles and also in terms of image, identity, creation of social living spaces, urban density, accessibility, development and even bringing populations together. The second goal is to build interregional partnerships and cooperation spaces to help cities, in both the South and North, which are struggling to become cities in their own right, and also to help those facing environmental, reputation, social cohesion and ageing infrastructure problems to overcome these difficulties.

The partners and contributors in the project are the Marne-la-Vallée/ Val Maubuée agglomeration community in France, the NGO Touiza Solidarité, the European New Towns and Pilot Cities Platform (ENTP), Egypt's Housing and Building National Research Centre (HBRC), the public agency in charge of managing the new city of Sidi Abdellah in Algeria (EPA-ANSA), and the al Omrane-Chrafate Holding in Morocco. The project, initially scheduled to run for three years, was extended by a year following the Egyptian revolution to take account of the ensuing changes, not only in policy terms but also in terms of priorities regarding housing, the involvement of civil society, and town and country planning.

---

3 http://www.ciudad-programme.eu/.

The CIUDAD (cooperation in urban development and dialogue) programme is part of the European Union's development cooperation and is funded through the ENPI – European Neighbourhood and Partnership Instrument. Inspired by experiences gained in the Med'Act and Med-Pact programmes, CIUDAD supports direct cooperation projects between local authorities and civil societies in European countries and 16 border countries.[4]

In December 2008, EuropeAid,[5] the European Commission's Directorate-General for development cooperation, launched the first call for proposals for the CIUDAD programme. The main objective of the programme is to promote mutual understanding, dialogue and cooperation among local players in the EU and neighbouring partner countries, by supporting efforts to enhance capacities in order to modernise and strengthen local and regional governments. The three thematic priorities are: environmental sustainability and energy efficiency; economic development and reduction of social disparities; good governance and sustainable urban development planning.

Put simply, the CIUDAD programme is one of the rare European programmes that promotes and subsidises work on decentralised cooperation in southern Mediterranean countries and that imposes conditions in terms of expected results and evaluations, hence its significance. It also requires those involved in the project to strike a balance between policy decisions, local priorities and regional problems and to develop open cooperation between cities of countries whose borders are not open.

Furthermore, while the "urban" phenomenon has been fashionable since the 1970s as a topic of study and object of policies, and recent years have seen mass urbanisation worldwide come under the spotlight, it should be noted that the European Union has no explicit authority when it comes to urban development, however local authorities and cities are increasingly being seen as players in European discussions and in implementation of policies (such as regional policy or through structural funds).

---

4 15 of the 16 countries from the so-called neighbourhood area (including partner countries in the Barcelona Process with the exception of Libya: Algeria, Egypt, Israel, Jordan, Lebanon, Morocco, Syria, Palestine, Tunisia, plus countries in the East of Europe: Armenia, Azerbaijan, Byelorussia, Georgia, Moldova, Ukraine), plus the Russian Federation.

5 Now "DEVCO", a new Directorate-General responsible for developing European development policies and supplying EU aid worldwide via projects and programmes. It combines two former Directorates-General: DG Development and DG EuropeAid. Source: http://ec.europa.eu/europeaid/who/index_en.htm.

The New Medina project has proposed a methodology built on exchanges and the organisation of events and expert panels. This initiative has allowed several seminars and expert panels to be held on topics as diverse as development techniques, governance, eco-construction and land management. These panels and seminars have made it possible to structure the exchanges and benefit from knowledge about the new cities, at the same time proposing practical recommendations for the project's partners. As regards the expected impacts, pilot studies and actions have also contributed to the project by putting forward practical solutions such as building a tramway between Algiers and Sidi Abdellah, creating a roadmap for Chrafate's development and encouraging training for Egyptian leaders in eco-construction led by the HBRC. Studies into major changes in regions and the expected impact on local populations are also coordinated by the NGO Touiza Solidarité.

This joint work highlights the results of the New Medina project, based on exchanges and cooperation between cities and associations from different countries in the North and South. The authors are of diverse origins in terms of both location and profession (sociology, town planning, local development, geography, political science) and offer different perspectives. Some articles set out an interconnected vision of new cities (R. Sidi Boumedine, P.A. Barthel, P. Gaborit), while others contribute with contextual aspects (C. Horn on town planning, L. Lahery on sustainable development). There are also articles that endeavour to provide solutions for building cities that are more sustainable (H. Farouh, D. Ruggeri).

Through the pages, as throughout the New Medina project, certain elements have appeared, as though guiding threads: the marking in time of new cities, notably European, now several decades old, the difficulty of linking together such complex projects in different stages between the short term and long term, and the need to adjust aims and constantly update them according to different constraints relating to governance, resources and implementation. The need to take account of the ageing of these brand new cities in terms of urban management is also a question that counterbalances the "ideal new city traveller" who, as Deni Ruggeri mentions, is more than two hundred years old. Finally, the question of ranking priorities has gone global: the creation of more cities requires costly programmes that can come into conflict with other needs such as resolving problems connected with informal housing or access to quality services and infrastructures in already urbanised zones.

The depth and interest of this book can be found in its mix of perspectives.

About New Towns

The Contradiction of New Medinas

Pilot Projects, Complex Territories, and other Issues of Sustainability

Pascaline GABORIT

Director of the European New Town Platform/Pilot Cities, ENTP

> The history of the Mediterranean has not grown along the lines of the Imperial Empires. It is fundamentally a network of Towns or rather of cities. The History of the Mediterranean is also the History of Mediterranean Cities.[6]

Introduction

In the 20th century, one attempted solution to real, imaginary and perceived problems of large cities has been the construction of new urban areas, intended to redistribute population and activities from the main cities and reshape the territories. It is in this context that New Towns have been created.

New Towns are usually understood as local authorities built from scratch following a master plan. This planned approach is opposed to organic growth when the towns develop and transform by themselves.[7] New Towns have not appeared but have been built to answer to demographic growth and urban pressure, to provide housing, areas for living and alleviate traffic congestion of larger neighbouring cities. In Southern Mediterranean countries some new towns are even being designated in desert areas and are considered as technical challenges to conquer these territories for human settlements, with questions remaining on sustainability, quality of life for future inhabitants and feasibility of these projects in the long run.

---

6 Balbo, M. (ed.), *Medinas 2030, Scénarios et stratégies*, Paris, L'Harmattan, 2010, p. 26.

7 See: INTI definition of New Towns-ENTP definition, http://www.pilotcities.eu.

Across different countries, and although the situation differs in many ways, New Towns share common challenges and features in their process: to anticipate the urban growth and planning for the next decades, to create new spaces, new areas of living. This involves solutions in a rapid amount of time raising the issues of prospective, risk analysis, finance, the disruption for the surrounding region and communities. In this article, we will approach the analysis of New Towns, as if they were separate and independent entities in a risky comparative perspective and describe the different development of New Towns both in Europe and in Southern Mediterranean countries: Algeria, Morocco and Egypt.[8]

In particular the New Towns of Chrafate close to Tangier, Sidi Abdellah and Boughezoul (Algeria) and New Fayoum or 6th of October (Egypt) are presented. We will base our findings and conclusions on the reports, panels, recommendations and events of the New Medina project. This publication is indeed the result of a collective work and thinking with the partners and experts of the project who have presented recommendations to the different towns.[9] Many questions emerge in the current debates: Is it still relevant to build New Towns? Should the national governments not for instance invest funds in priority to solve the problems of informal settlements or the renovation of old city centres? The aim of this book is not to answer to these questions as much to detail the findings of the project, the trends and perspectives on existing New Towns as crossroads for more understanding on sustainable urban development.

New Towns have been created in Europe after World War II and are currently being designed in Egypt, Algeria and Morocco. But do we only assist in a different process where we would oppose older New Towns (like in Europe) and New Towns to be built? Are we referring to the same "New Towns" concept or is it different? Have we simply "copy pasted" the principle or is a new idea (or New Towns' mindset)[10] emerging? Is it not time to make also the "New Towns" concept evolve towards a "pilot cities" idea where scales are smaller (planned area by

---

8 New Towns are planned in Egypt (22) in 3 different generations with a national program that started with Nasserian New Towns. In Morocco the New towns projects include the town of Tamansourt close to Marrakech, Tamesna nearby Rabat, Chrafate in the outskirts of Tangier, Lakhyata close to Casablanca, and the future Eco-town in the region of Benguerir). In Algeria 5 new towns are being planned: Sidi Abdellah and Bouinan in the periphery of Algier, Boughezoul, Tafilelt (in the north of Sahara) and the New Town of Hassi Messaoud.

9 The comparative perspective gives some elements to analyze the New Towns' development, but is setting up limits as each country and framework present different patterns.

10 Gaborit, P., "Reinventing the New Towns Mindset", *Nova Terra*, Delft, September, 2005.

area) and where it is acknowledged that some investments are experimental (for instance, to test new building techniques)?[11] Will Eco-cities and Eco-areas replace the New Towns Concepts?

We need to acknowledge that in the past decades the global context has changed for public urban policies: public led initiatives are scarce. Public resources are difficult to mobilize and the long-term investments needed within a New Town are difficult to justify. But it is astonishing to realize, that issues identified in existing New Towns (lack of public transport, absence of participatory approach, low densities) remain identical in the New towns to be built as a homogeneous, often not flexible, urban project to answer to urgent demographic needs and demands for housing.

Therefore the paper will raise some elements of analysis drawn from the different panels and papers, all acknowledging that New Towns can be a controversial topic and a risky venture, with the necessity for planners and decision makers to think ahead in order to avoid the following risk: the creation of towns remaining empty with many "houses without people", in countries where informal settlements are still an issue.

I. The Birth of New Towns and Why They Are Created

It is generally acknowledged that the origins of towns and cities are to be found back in the Neolithic agricultural revolution, and that the first cities have emerged in the "fertile croissant".[12] The surpluses of food could be exchanged what paved the way for the development of commercial exchanges.

In the history cities have indeed developed along commercial development and exchanges. The idea of New Towns creation promoted by Ebenezerd Howard[13] constitutes a disruption.

---

11 In the dictionary, a New Town, or planned community is a city, town or community that was designed from scratch and grew up following a master plan. The ENTP defines New Towns as "towns created in the periphery of large agglomerations, following a planning approach". In Europe, after World War II, the aim of the New Towns was to provide housing to the cities (part of them were destroyed during the war, others were expanding industrial sites). At that time, these New Towns were mainly dependent on the larger cities for the provision of services, leisure, cultural events, retail, etc. In the UK, the Department of the Environment decided on the creation of the New Town program, where 21 New Towns have been established since 1946. Only in the London area the plan was to relocate around two million people. In other countries the initial plans were less ambitious. In the French case, five New Towns were designed around Paris.

12 Currently Syria, Iraq, Jordan, Israel and Palestine.

13 Howard E., "*The garden cities of tomorrow. A peaceful path to Real Reform*", 1902.

Nowadays "*As globalization increasingly extends urban forms across the world and integrate existing cities into vast urbanized systems of communication, transnational flows of finance, commodities, labor, images and ideas, the idea of the city as an organism, defined by an internally coherent civic life and structured by clear relationships to the region, nation, and wider world appears obsolete.*"[14] The initial projects of the creation of New towns however; seem to answer to a number of consistent reasons.

In a context of demographic and urban explosion and urban migrations to capital cities, appears a need to improve the living environment, avoid informal settlements (slums in the outskirts or at the edge of the cities) and to organize the urban development.

In this context, New Towns can appear as solutions. But we cannot only restrict the decisions of their creation by their functions. They are also planned as "modern cities", as a symbol of modernity and achievements by national governments, often in a rather top down approach. It is also interesting to notice that they integrate a vision of the territory as a "land to be planned and transformed" from scratch, an area to conquer and where well off or middle class inhabitants will be invited to settle later on. In this sense, New Towns are not "new communities" that would integrate spontaneously inhabitants' needs. They are rather large projects of development launched by national governments or private developers to transform an empty area or a rural area into a liveable town. They develop progressively into large scale projects with targets and deadlines in terms of housing units, development, provision of infrastructure, etc.

New Towns are neither sole "projects" of governments or developers only. They require the involvement of a range of different stakeholders, including existing local entities, private sectors, the public authorities in charge of infrastructure, as well as large teams of urban planners, architects and engineers that are involved in the town's development from its design to its realization.

At the moment of their birth, New Towns appear as alternatives to growing suburbs where informal settlements prevail, informal settlements often being associated with location of protest and contestation towards the national state and the governments.

In Western Europe, New Towns have mainly been created around large cities like London, Paris or Amsterdam after the World War II, in a time of construction boom in the future with the idea that governments would only provide "welfare" for their citizens, but also a brand new

---

14 Gyan Prakash, *The spaces of the Modern city: Imaginaries, Politics and Everyday Life*, Princeton University Press, 2008.

environment. The prevailing impressions at that time were that resources such as oil would be unlimited, that demography would remain unchanged; that growth would not drop and that we would follow a steady evolution towards progress.

In Southern Mediterranean countries, this "functional" vision of New Towns is not absent from the national territorial master plans, with the objectives to rebalance the national territories from highly populated littoral areas to more deserted towns. In Egypt and Algeria in particular, the population is highly concentrated in the capital cities: Cairo and Algiers. As a national vision of the territory, New Towns are then developed to alleviate the traffic and to compensate the overweight of these large metropolises.

When they are not led by private developers, New Towns are being implemented in public national programs. In the idea of a Top Down approach; decisions are taken at different levels, with injunctions to the lower entities (ministries, local entities, and territorial agencies) to find innovative solutions to "make it happen". The idea to "start with the Community" and local needs is often absent from the step of the conception of the city's conception.

A. Decisions, Contexts and Booming Population

One of the reasons for governments to create New Towns is a booming demography, the congestion of large cities, or, as in the cases of Algeria and Egypt, the concentration of the population in a relatively small part of the territory, and the idea that these new towns will contribute to the development of these countries.

1. Territories and Sense of Emergency

The comparison between European New Towns and Southern Mediterranean New Towns is rather difficult for several reasons. One of these reasons like in the case of Asian developed New Towns is the scale and sense of emergency: the needs in terms of housing provision, sanitation, and relocation of population are without a doubt more important in Egypt, for instance where informal settlements are very important by comparison to post-war Europe. The climate and territory is also different.

Whereas in European countries like for instance in the Netherlands, land was reclaimed on water to create New Towns like Almere and Lelystad, the Egyptian New Towns and the Algerian New Town of Boughezoul are being built in the desert. This creates specific territorial constraints for the construction of housing units (as described in the

"Egyptian referential for sustainable cities"),[15] in terms of heat prevention, shadows, water provisions, etc. The impacts on the territory are also very different when the New Town is being built to alleviate the congestion of the main metropolis (like French New Towns) or when they are built to balance the national territory by creating a self-contained community far from the capital (like it was the case of Boughezoul). Deni Ruggeri makes a clear distinction between both models: when the synergetic model of New Towns when they are created as "in complementarity with the metropolis" or the model "in competition" when the New designed town is an independent town or a future large city.

2. Urban Migrations and Territorial Balance

The increase of demography is often associated with migrations to urban areas: In 2008 it was expected that Egypt's population would likely increase by 27 million inhabitants to reach over 100 million. Accommodating this large amount of population in a short amount of time is a major challenge for the different governments, with a necessity to provide affordable housing and urban facilities.

In Egypt only, the two primary cities: Cairo and Alexandria comprise 43% of the total urban population.[16] This concentration of population in Cairo in particular is associated with the prevalence of informal settlements,[17] urban poverty, unemployment, environmental degradation, and marginalized populations.

These informal areas of Cairo were mainly forgotten by national project's development and instead managed by self-organized communities, more prompt to be active in contestation.[18] The rest of the country is mainly covered with deserts, what made the governments decide to claim the desert areas and to build New Towns as a part of the national development strategy.

---

[15] This referential had been prepared in September 2009 by experts for the Egyptian council on sustainable development for the national program with the help of the World Bank. This document is still actual (Goell, E. *et al.*, *Sustainable Cities in Egypt. Learning from Experience: Potentials and Preconditions for New Cities in Desert Areas*, The Egyptian Cabinet, Information and Decision Support Center, Center for Future Studies, 2009).

[16] General Organization for Physical Planning, June 1998.

[17] According to the World Bank 2008, 16 million urban inhabitants live today in informal and squatter settlements.

[18] Dorman, W.J., The Politics of Neglect: The Egyptian State in Cairo 1974-98, Thesis, School of Oriental and African Studies, University of London, 2007.

Figure 1. The generation of new Cities in Egypt

NUCA

In Algeria, the national strategic document: the SNAT also analysis a necessity to relocate one part of the population currently concentrated on the coastal area further in the mainland. The same idea led to the designation of Boughezoul, a previous project to build a city in the semi-desert highlands as possible future capital of the Algerian state. The construction has started although the proposal to relocate the capital to Boughezoul has been temporarily forgotten, but is coming back from time to time in a cyclic manner.

The New Town of Boughezoul in Algeria, 2011

3. A Real Programming!

The New Towns policy in Egypt has been an ambitious policy since the Nasser regime designation of Nasserian communities). In the 1970s and 1980s several New Towns have been built around Cairo such as the towns of 6th of October, and 10th of Ramadan. The graph above shows the different generations of New Towns, although it is not very clear whether there are significant differences in the way the different towns have been planned. Already in 1977, not less than 19 New Towns were planned. In Egypt, indeed, New Towns were considered one of the best ways to develop the country's territory towards the desert, as symbols of modernity and "towns of the future" for the state and government accused of neglecting "informal areas".[19]

Most of these Egyptian New Towns are extensions of the desert. Initially planned to provide housing and accommodation, over the years they have become, for some of them, private sector led initiatives, to the benefits of the wealthier communities, when they have not remained deserted by the population even if they could be at the first steps in their history.

Egyptian New Town of New Fayoum

New Towns also reflect the ambitions of the different governments to achieve progress, modernize, and create a more balanced national territory reflecting sometimes not so realistic ambitions. For instance the President Mubarak declared in 1996: "The conquest of the desert is no longer a slogan or a dream but a necessity dictated by the spiral of the

[19] *Ibid.*

population growth. What is required is not a token exodus into the desert but a complete reconsideration of population throughout the country".[20]

In Algeria, the designation of the New Towns has not been made in the framework of a program for New Towns, but rather as an investment for balancing the national territory, with the initial idea that Boughezoul could develop into a new capital city for the country. In Morocco, a program was set up to create 5 New Towns, all of them with a different layout and function. There was not, for instance, any law regulating the creation of the New Towns.

It also has to be acknowledged that the decisions from the governments and the public programs to create New Towns are decisive for the location and development of these towns, especially when the location is decided by the highest authorities. The choice of the location in particular has strong impacts for the development of the city, as there can be several risks such as flooding, wind corridor etc. Finally, other projects have emerged recently. In Morocco the project "Green City" eco development in the outskirts Benguerir will be named Mohammed VI City.[21]

4. Is a Law about New Towns Necessary?

Different frameworks and organizations have been set up to plan and develop New Towns in the different countries. In Egypt, the New Urban Community Authority: NUCA was created in 1979 and given power to develop the New Towns and manage them. The NUCA represents the centralized state's power on the territory of the Egyptian New Towns.[22] In Algeria, the New Towns have been created on the French model with the creation of public agencies in charge of the New Towns' development. The actions of these agencies are detailed in a law dated from 2008.[23] Different New Towns have been designed in Algeria; Boughezoul, Sidi Abdellah and Bouinan. Indeed for the designation of each of these New Towns, the law describes the location, perimeter and surface, as well as their economic[24] specialization: Boughezoul is being designed as potential future capital, and to host a centre on renewable energies (the first centre to be welcomed would be a research cluster on

20 Goell, E. *et al.*, *Sustainable Cities in Egypt. Learning from Experience. Potentials and Preconditions for New Cities in Desert Areas*, The Egyptian Cabinet, Information and Decision Support Center, Center for Future Studies, 2009.

21 Press release "Morocco is building a Mega 'Green City' named after King Mohammed VI in Architecture and Urban", November 12th, 2012.

22 In cooperation with the General Organization for Physical Planning G.O.P.P. and the Housing Research and Building Centre H.B.R.C.

23 Related to the creation of New Towns and their development, in 'Law', No. 02-08, 8th May 2002.

24 With an incubator and a multi-businesses building in Sidi Abdellah.

energies). Sidi Abdellah is planned to welcome incubators and clusters on new technologies.

In Morocco, discussions have taken place for the adoption of a law about New Towns.[25] This law could have had the advantage to clarify the procedures for the development. It is now indeed not very clear which competences would be the responsibility of the developer (Al Omrane) and the municipality (Joumaa in the case of Chrafate). As this law has not been approved yet, the general law on urban planning is providing the framework for the development of New Towns.[26] In the case study of Chrafate, the developer which is also the promoter, acts in a more general framework and does not benefit from the specific competences that this law would have granted to him (among others as examples: the organization would then act on behalf of the municipality and its president would act on behalf of the municipality).[27]

In Morocco more in general, several reports question the means given to municipalities to perform their tasks.[28] In the framework of some towns, there have been some doubts whether the municipality would be able to undertake the management of the New Towns, once this one has been created.[29] The existence of a law or national program for New Towns depends from country to another. The existence of such legislation though facilitates the procedures. There are many reasons to for public led New Towns to emerge: booming demography and the needs of population in terms of housing. In a context of liberalized economy, will the private sector replace these public programs?

B. Gated Communities

As we have already stated, the development of the New Towns is different depending on the context and time, the country, and the climate. But the global context has also changed and evolved. After World War II, it was well accepted that the national state would provide subsidies and expenditure to develop the facilities, and develop housing for their citizens. Although the land management was used to make sure to compensate the development prices by sale prices, public governments were in charge of these New towns developments (except in the case of British private led garden cities).[30] Nowadays, the global context pro-

25 New Medina Panel, Chrafate, 2010.

26 Related to urban planning, 'Loi', No. 12-90, 15th July 1992.

27 Boumedine, R.S., *Étude comparative*, Panel of Marseille, p. 8.

28 Le Tellie, J., F. Debbi and L. Amzil, "La mobilité urbaine dans l'agglomération de Tanger; évolution et perspectives", Blue Plan, October 2009.

29 New Medina Panel of Chrafate, 2010.

30 This is of course very different in the American cities like Irvine.

motes a certain liberalization of economy, the scarcity of public expenditure, the development of public-private partnerships, and this forces government to ask for a quick return on investment for their national policies.

Although the different Southern Mediterranean countries did not follow all the constraints asked by new global order, there has been a shift from public state led initiative to more private led initiative that arrived with a series of impacts for the development of the New towns.[31]

In Egypt, after the first generation of New Towns, private developers emerged, breaking with the strong state tradition since Nasser, and created facilities and living spaces of high standing, more profitable on the short term. The architectural design, often inspired by modern global cities like Dubai, is not really questioned. Among others, Mona Abaza in her article "Critical commentary: Cairo's downtown, imagined Dubaisation or Nostalgia"[32] states that territorial space in Cairo and in its surrounding is more and more segregated "by the development of islands of luxury". She suggests that planners and decision makers do not propose alternatives between "dubaization" and "Nostalgia" or the old city.

In Egypt, land along the desert is sold at very low prices by the state, which owns the land. Private developers can invest and create new spaces, compounds or "gated communities", fragmenting space.[33] The allocation of land in these Egyptian cities has also been very much questioned.[34] Land prices for private entities were not very high infrastructure costs were not duly integrated in the prices of sales which proved to be very convenient for promoters. This encouraged the purchase of lands without obligations to build, and allowed the purchase of land that remained empty, although the legal framework foresees that the territory can be repossessed after a certain amount of time if it is empty. Speculation and empty houses are indeed the risks of a rapid urban development.

This can explain that decades after the creation of New Towns only a really small percentage of the population is benefitting from them: the wealthiest by offering expensive housing, leisure complexes, shopping

---

31 Barthel, P.A., "Premiers quartiers urbains durables dans les pays arabes: enseignement sur une génération spontanée", in *Espaces et Sociétés*, No. 147, CAIRN, 2011, pp. 99-115.

32 Abaza, M., "Critical Commentary. Cairo's Downtown. Imagined Dubaisation or Nostalgia", in *Urban Studies*, No. 48, May 2011, p. 1075.

33 Florin, B., "Des cités nassériennes aux villes nouvelles du désert: la fin du logement social?", in Battesti V. and F. Ireton (eds.), *L'Égypte au présent. Inventaire d'une société avant révolution*, May 2011, p. 142.

34 Yousry, S.A., *Case Study: Urban Transformations in New Cities in Egypt, 6th of October City*, Cairo, 2010.

malls etc. "Gated communities" are also common: In 6th of October, not less than 50 "Gated communities" are registered. However few of them are really known as exemplary models for development. But if "gated communities" are highly questionable in terms of segregation of which they become the symbol; we could almost say that they appear more enviable than the "empty cities", ghost towns that remain empty and "abandoned" dormitory cities that are "towns without a soul".

Figure 2. Characteristics of the different New Towns

New Town	New Fayoum	Sidi Abdellah	Ch'Rafate
Location	15 km from Fayoum 100 km South East from Cairo	25 km South West from Algier	20 km South West of Tangier, 25 km from Tetouan and 25 from Tangier Med Harbour
Creation	Decree 2000-3rd generation of New Towns launched in 1999	Decision 2004 after the law 2000 on the creation of New Towns	2009
Surface	650 ha	7,000 ha global territory-Green belt 4,000 ha and urban development 3,000 ha	770 ha
Target population	120,000 inhabitants	200,000 inhabitants and 80,000 jobs	150,000 inhabit-ants
Housing units	40,000	50,000	30,000
Forecast expendi-ture for develop-ment	-	-	206 million Euros
Current population	Around 100	Around 48,000	150 families – 3 villages (douars)
Climate	Desertic	Coastal-rural agricultural territory	Coastal-strong winds

C. A First State of the Art: The Risk of Emptiness and Scattering

It is estimated that a lot of public funds have been spent for the planning of New Towns in Egypt only (22% of the budget for the ministry of housing between 1998 and 2002)[35] for very few percent of the population. It is estimated that in 2008, only 3% of the inhabitants from the Cairo metropolis[36] do live in New Towns. This trend prevails even if New Towns such as 6th of October represent an important part of jobs and employment opportunities.

The absence of public transport to and from the New Towns, as well as the absence of large roads (together with the high prices for housing) may explain this "desertion" on New Towns even among the higher income groups.[37]

Another difficulty faced by Egyptian Towns and its managing organization like NUCA, is that New Towns may appear as "icons" of the previous regime of Hosni Mubarak, and it may be difficult to carry on their development after the Revolution. This is also what Dorman mentions in his PhD thesis *The Politics of Neglect*: "Although informal areas had been abandoned by the public policies during the previous Regime of Hosni Moubarak, huge urban development projects were developed to benefit the corrupted elite, reinforcing the politics of patronage".[38]

On the contrary, the informal housing sector in which 60% of the population is lives has been "forgotten" from and neglected by the public policies. This informal housing sector is important and allows the population to remain close to their family and communities, as well as to build their own houses.

Emptiness, scattering and urban sprawl are the main pitfalls for the development of New Towns.

There are many consistent reasons for the birth of a New Town and the initial project is often consistent because of the objectives, the ambitious measures implemented and the transformation of territories. The initial project will however adapt and evolve in the length of time and take into account the different urban times: what means to combine the short and long run.

35 Florin, B., *op. cit.*, p. 142.

36 That currently counts around 16 million inhabitants.

37 Metropolis Conference, October 6th, March 2008.

38 Dorman, W.J., "The Politics of Neglect: The Egyptian State in Cairo 1974-1998", PhD Thesis university of London.

City of New Fayoum

Apart from these questions, the locations of the New Towns, like the one of New Fayoum in the desert, are often questionable. This was one of the conclusions of the experts who met in Fayoum in 2012.[39] The decision to build a New Town from scratch without real consultation of inhabitants and of their needs may lead to the risk of failure. The question of materials used in the construction of the first buildings without consideration for the climate (no cooling system for the buildings, resting on a generalization of air conditioning) is also difficult to assess. Finally, the mono-functional aspect of housing areas with limited provision of services is also a threat the development of these New Towns.

II. The time Cycle of the Town: Short and Long Term Visions

"*The great obsession of the nineteenth century was, as we know, history: with its themes of development and suspension, of crisis and cycle, themes of ever accumulating past.*"[40] Cities evolve in a historic timeframe as shown by the different landmarks and monuments that provide them with continuity for the historical towns. The time of the city depends of this vision of continuity, extension of space and of long term. In the case of New Towns, the articulation between the short-term

[39] New Medina, Panel of Experts, 2012.

[40] Foucault, M., "Des autres espaces", *Diacritics* 16, 1986, p. 22.

management and the long term becomes extremely complex. The New Towns indeed the different stakeholders to shorten the time necessary for their decisions, the implementation of procedures and adaptation in order for them to be created in a short run, with a view of a long term transition.

The New Medina project gave to the different experts the opportunity to think further and discuss about the development of New Towns when they are at their early stage and need to anticipate their growth towards mature cities. The project identified tools and recommendations that can be described below and discussed further in the different stages of New Towns' development. For this purpose, the integration of the New Town in its surrounding territory, region and community is of high importance.

A. Land Management

Land management is at the core of the Towns' development: "Land management is the raw material of urban planning and is therefore to be considered as being strategic".[41]

Land management is, in the context of the creation of New Towns the main tool for public governments to succeed in the development and avoid speculation that occurs when the land prices are initially very low and increases when the objectives of the master plan and New Towns' development are known.

This is why the British New Towns development corporations, as well as the French *Etablissements publics*, acquired a land reserve that was used to keep the prices of developed land at a reasonable price. The land lots were sold to promoters once it has been equipped with the relevant infrastructure like sewage, water and energy provision, the sales price including the costs of equipment. It has prevented the speculation in the zones designated for the New Town's development.

In Egypt, the system used is similar in theory but the used procedures have encouraged quick sales of land plots without long term vision, which has created problems of land fragmentation. In the case of Egypt for instance, the government has sold the equipped land at very convenient and low prices to promoters and individuals. This resulted in quick sales of land used for speculation without intention to build the housing units. As a consequence, a certain portion of areas sold for housing have remained empty despite the possibility for the NUCA to repossess the land if the houses have not been built after a certain amount of time (usually 3 years).

41 Soladaïni, P.L., EPF, Panel of Marseille, March 2012.

That being said however, the developer of the New Town will need to find a balance between an efficient implementation of the master plan and the constraint of flexibility. Selling land plots quickly can indeed secure the developer for a quick return of investment, whereas the master plan consistency would require a more long-term consideration, investment and advance of funds.[42]

In order to find a rhythm in the development of the New Towns, balancing flexibility and consistency, some New Towns have proposed a system of zoning. Under this system it is expected to deliver the towns in different areas at different points in time of the New Towns' development. Since the New Towns are being built in a very long term (30 or 40 years ahead), it is difficult to avoid a phasing in the development. However, the coupling between the temporal phasing and the spatial zoning has been considered as a failure in the past,[43] as it creates the risk of disconnected functional areas as opposed to a consistent integrated territory. The question of having a "critical mass of inhabitants" before starting to further build the town may well be one of the criteria of success for the development of zoning.[44] In the different phasing moments of the New Towns, it is necessary to achieve a certain liveability of the area created to allow a smooth development of the town.

B. Urban Forms, Landscapes and Quality

The New Town is confronted in its realization and development to the territory's patterns. It needs to be integrated in the current landscape: the natural landscape on the one hand and the built landscape on the other hand.

The integration of the New Towns in its landscape is a very important pattern for its development. Some authors like Deni Ruggeri indeed suggest in the case of Irvine that the natural landscape is more important for the perception of inhabitants on their place of living, than the architectural landscapes. Moreover some theories argue that well managed urban and rural landscapes could significantly increase the resilience and the adaptive capacity of the territory towards the expected negative impacts of urbanization.[45] These researches and theories show how important it is, not only to build from scratch but also to integrate

---

42 Except if a system is set up to finance the development of the New Towns with long term bank loans with low interest rates, that would avoid the endless search of quick returns on investment which favor speculation.

43 Gaillard M., Panel of Chrafate, 2010.

44 Panel of Experts New Fayoum, 2012.

45 Hudekova, Z., U. Ferber and E. Guglielmetti, "Recommendations Euroscapes project", http://www.euroscapes-eu.org, 2012.

the different aspects like geology, water, vegetation existing and imported, blue and green spaces, and the use of elevations. "The conception of the New Towns needs to be dynamic, taking stock from existing patterns and not developed from scratch."[46]

The New Towns' developers are indeed in charge of all aspects of a development: creation of urban forms, architecture, transport axes, and a city's image, while at the same time developing procedures for the realization of the master plan.

The question of quality also has been discussed among the participants of the New Medina panel. The adoption of a charter by the developer of Chrafate has been considered as a key point for the quality of the future town to be. It, among others, describes the constraints in terms of comfort, aesthetics, and energy efficiency.

New Town of Sidi Abdellah on a hilly landscape, 2011

[46] Boumedine R.S., Panel of Marseille 2011.

Construction site of the town of Ch'rafate on a hilly landscape close to a field of wind turbines

C. Institutions and Governance

There has been a lot of literature about multi-level governance, and the increasing role that local authorities are playing. Governance is defined as the art of public actors to work together.[47] In the case of New Towns, the challenge of governance is twofold: to find temporary management organizations to be able to develop the New Towns before the transition, and to prepare the New Towns for their future management as any local authority.[48] This double objective is very central to the construction of New Towns.

1. Governance During the Development of the New Town

At the moment of their creation, New Towns do not have yet a critical number of residents as other municipalities, and do not have a local representative council in countries where this is usually the case. In some cases, the initial number of inhabitants is very low, and the competent rural municipalities do not always benefit from a lot of means and capacities to manage the future New Town.[49]

Before a "transition" period it is accepted that one organization representing a public or public/private authority is in charge of the New

47 Güntner, S., Multilevel Governance, a Brief look at the Current debates in EUKN Multilevel governance, the art of working together, 2011.

48 Panel of Marseille, March 2012, report p. 13.

49 Panel of Experts in Chrafate, 2010.

Town's development. In the case of Morocco, the holding al Omrane is in charge of the development of several New Towns. The New Towns of Sidi Abdellah and Boughezoul in Algeria are developed by an *Etablissement Public D'Aménagement* which is a public entity representing the national state. In Egypt, the New Urban Community Authority NUCA is in charge of the New Towns' design, development, and management on behalf of the national state.

The previously mentioned organizations that are acting as developers of the New Towns, before the period of transition (that can take several decades, like in the French case) are facing several challenges:

– Find a legal frame to build quickly and provide facilities to the inhabitants in a short amount of time. Indeed, if urban planning general rules are to be applied, the time necessary to make the first areas of the New Town livable may be too long.

– Preparing the future municipality to manage the New Town. This is a challenge in itself to make sure that the municipality that will be in charge of the future New Town will have enough capacities, staff, legal authorizations to be able to continue the New Towns development and capacities

– Clarify procedures and responsibilities: in the case of the New Town of Chrafate for instance, the experts of the New Medina project have to provide some support to identify possible ways of joint work between the different stakeholders in charge of the New Town's development and future management. This only necessitated one study, as it was unclear for the developer how to best develop and implement the master plan while integrating the New Towns in its current administrative territory.[50]

– Find enough financial means to ensure the provision of facilities for inhabitants (equipment for the housing areas, energy and water provision) but also public schools, transport, parks, leisure and cultural facilities that are expected by the inhabitants.

The involvement and consultation of future inhabitants is a key success factors for the development of the New Town.[51] However, in the history it is difficult to find real consultation and citizens' involvement in the planning and development phase of the New Town. In the case of European Towns, for instance, consultation and citizens' direct involvement began quite late in time after the transition and the recogni-

50 New Medina project, Study on Governance for Chrafate.

51 Panel of Marseille, report p. 9.

tion of the existing New Towns as normal municipalities.[52] Nonetheless the involvement of population from the early stages presents several advantages:

- Development of a pioneer spirit among the first inhabitants of the New Town as it was the case in some of the British New Towns.[53]
- Base prospective on realistic trends, and avoid a top/down approach disconnected from reality for instance while evaluating the targets for future to be built housing units.
- Make the housing supply match the demand.
- Provide facilities that are really needed without loss of funds.

As it was said during the different panels of the project, the developer has very challenging tasks to consult, but also negotiate with the different stakeholders and decision makers on the territory.

2. *The Transition and Future Management*

The transition at the time of the creation of the New Towns appears to be a far, blurred horizon, almost like a target in itself reflecting a kind of "ending" of the New Town, as if a town would not be always evolving with several constraints of regeneration and aging of buildings. This is one of the lessons of the European New Town.

In the case of European New Towns, in some countries the transition has been rapid like in the UK, where the development corporations have been wounded up at the end of the 1970s and during the 1980s. In other countries like in France or in Finland the transition from the New Towns statutes to normal statutes has taken several decades.

Construction sites in Sidi Abdellah

[52] See: examples of consultation for instance the "political market" in Almere New Town.

[53] The New Towns of Stevenage and Harlow for instance.

D. The Image and Identity "Under Construction"

If it can be said that Towns can have an identity and image as an entity, this image and identity of New Town would be central to their development. In Europe, and a long time after their creation, it is generally recognized that New Towns suffer from a problem of image and identity.[54] Nonetheless this question is already present at the time of the New Town's development. The New Town's image is indeed considered as twofold: the real image of the New Town in construction, and the fictional image of the New Town to be after the master plan is completed.[55]

We should acknowledge the following opposition: New Towns are being built for the long term, but promoters sell its image of already completed town.[56]

This "fictional city" which is being presented to the promoters, future inhabitants, and regional population, is necessary to sell the town. Therefore the urban marketing of the New Town is one of the pillars of its success. In order to achieve a good marketing of the New Town, the experts of the different panels of the New Medina project have proposed several recommendations:

- Identify the target groups: the future inhabitants, the first residents.
- Agree on what the New Town "*should be and should not be*". In the different panels of the New Medina projects, the different experts and stakeholders have discussed that Chrafate should be considered as a good place of residence for the inhabitants of the whole region.

54 See project NEWTASC: New Towns Image and Identity; Gaborit, P., *New Towns. Image, Identities, Future Perspective*, Brussels, PIE Peter Lang, 2010.

55 Panel of Sidi Abdellah, Final Report, p. 12.

56 Boumedine R.S., Report of the Panel Sidi Abdellah, p. 1.

In the case of Sidi Abdellah the strategic positioning of the city as integrated in greater Algiers[57] or as a self-sufficient town has different impacts for its current image and development.

- Be innovative, and involve communication to and from inhabitants.
- Have a work of consultation with businesses and inhabitants to adapt the equipment to real needs.
- Integrate the public equipment like schools, research centres, and cultural facilities as a sales argument for the New Town. In the case of the city of 6th of October in Egypt for instance, the lack of schools is considered as a reason why the New Town is still under populated.[58]
- Develop communication support tools, like posters and DVDs.
- Explain the New Town to the inhabitants.

The choice of the architectural quality also has an impact on the future image of the town. In the remote example of Eastern European New Towns (like Prague 11), the housing areas often lacked architectural quality, not only because of the initial plans made by the planners, but because of the necessity to build quickly (part of the architectural concepts were abandoned by the constructors).[59] This is why, more recently, the town of Chrafate has decided for instance to have a charter for architectural quality that will apply to future constructions. The design of some of the New Towns proposes too often homogeneous blocks and housing estates without real recognizable design. Moreover, the "standardization" of New Towns is also at stake. The images that prevail of New Towns and half-empty cities question their models of development and how they are developed and proposed. Is it the modern, global city like Dubai, but without the means to develop it? Are there other models and what are they? Some of the experts at the panel of experts in Cairo therefore recommended to give up the Dubai model of city development".[60]

---

57 With proposals that Sidi Abdellah should be the "motor" of the Greater Algier, "*Sidi Abdellah: your future in the greater Algier*" or "*Sidi Abdellah: Algier's new start*". This strategic vision has also an impact on the mobility strategic plan and the transport connections between the capital city and the New Town.

58 Yousry, S.A., *op. cit.*

59 See Exhibition on Communist Post War Architecture, Prague City Hall 2012.

60 Boumedine, R.S., Panel of New Fayoum, 2012.

III. Towards Mature and Sustainable Towns?[61]

In order to become sustainable, New Towns, like any other towns, need to combine healthy economies, social communities, and environment for the inhabitants. However they have the additional challenge to achieve a certain number of inhabitants and density to be considered as complete and/or *mature towns*.[62] There is indeed a risk for New Towns to become dormitory or ghosts cities.

A. The Economy: Attractiveness and Innovation

Economy and exchanges lie historically at the heart of cities' development. The inclusion of "economy" as a main theme for New Towns' development is also a continuous priority. In the early start of their creation, the necessity to find "customers" or "buyers" or promoters for the housing units is central to the new towns' development. This first aspect, which gives a lot of importance for the marketing of the town, conditions the further development of the town.

But the question of "attracting" businesses on the territory remains a key point after the completion of the New Town. This can be easily explained because local authorities possibly get revenues from the present businesses (in the case of Finland for instance, or the previous French professional tax), but also because they generate employment and increase the job/housing ratio for the territory. In the case of British New Towns in addition, the businesses present on the territory offer facilities such as trainings to the municipality.[63] New Towns in Europe have offered facilities to the private sector with the creation of business parks (Sénart, Basildon), the development of free public transport to business areas (Capelle aan den Ijssel), the new trend of creating clusters (Marne-la-Vallée/Val Maubuée) and the quite massive generalization of having a department within the municipality working for economic development and partnerships.

We, however, notice a difference between the New Towns that host a university or highly well-known research centre (French New Towns/ Harlow campus/Louvain-la-Neuve University) and the others (Dutch New Towns in particular did not benefit from having a university on their territory as the traveling distances to the large cities for students

---

61 The title of the joint vision document of Dutch New Towns is "New Towns worden grote steden".

62 The ENTP has organized several events on the theme "Are New Towns mature towns?"

63 See the conclusions of the NEWTASC project: "*New Towns image and identity*" and ENTP general Assembly 2012 "*How to engage the private sector in urban regeneration*".

were not considered as too important). This absence of research centres led to some difficulties for some towns to transform the economic zones into pilot areas like clusters. There is a current debate within some Dutch New Towns however, to create more creative clusters, and to transform empty offices into housing areas.[64]

In the case of the New Medina partners, the projects in Algeria in particular show a strategy for economic development with an emphasis on research and innovation, both in Bourghezoul (green energies) and Sidi Abdellah (new technologies). In Morocco, a "Mega Green City" project has emerged. This project led by a private university is located 70 kilometres north of Marrakech, and plans to integrate eco areas, biodiversity and planting of trees allied with the most modern techniques. In all the New Towns, there is a wish to develop business areas, with incubators (see picture below) and offices. In the case of Chrafate nonetheless, one important part of the business activity will be developed within the Tangier Med Harbour and there are therefore attempts to find solution for this New Town not only to become a suburban area of Tangier but rather an area perceived as dynamic and integrated within a wider territory.[65]

Among the aspects that make a new town perceived as dynamic or not, the presence or lack of retail, shops and restaurants makes a big difference among the New Towns. The absence of shopping areas, markets and small retail gives really easily an impression of "emptiness" that is often mentioned (should it be right or wrong) as "the absence of public space". In the case of European New Towns the development of gigantic shopping malls (Sénart, Hässelby-Vällingby, and Vantaa) located in the centre of the town also reinforces this impression of "emptiness" in the surroundings.

---

64 Project Creanov and discussion with the municipality of Zoetermeer.

65 The reports and deliverables of the New Medina project proposed among others a partnership with the harbour "Tangier Med" to develop further the new town.

Business incubator: New Town of Sidi Abdellah

B. Mobility

Mobility comes after accessibility. It is the possibility to access one point or the other when you are inside of the town. Accessibility is the itinerary to reach the town from another location. This accessibility to a town and city is one of the key factors of its development. The decisions on the location of the New Towns are often taken in function of their accessibility, the territory and the liveability of the areas. However, it sometimes happens that the proposed accessibility scheme depends mainly on road access without provision for public transport (soft rail, metro, tram, train) and regardless of future congestion problems.

The Southern Mediterranean New Towns did not really seem to take these questions into account, and the patterns for transport do not appear to be sustainable in the case that inhabitants need to commute to the surrounding cities to find employment.

Mobility comes in a second stage of development and it is generally acknowledged that it depends on the "culture" of the population: pedestrian access, bicycle paths, use of cars, shared cars, minibuses, public transport are different options for transport that highly differ from one region to another because of the culture, demography, development and also climate. It is difficult to draw general remarks on mobility as congestion seems to be a problem almost anywhere in the world where the population is dense. There are few projects of transport that could be replicated from one place to another without further consideration of the use, patterns, and communities.

In particular, mobility is still a challenge in European New Towns. Several decades after their creation, the time was still not sufficient enough to develop an entire integrated transport system connecting all parts of the territory. The case of Val Maubuée, lead partner of the New Medina project, is very illustrative: the city is very well connected to Paris by the rapid Express Train RER that is now saturated and a service of local bus transport. However the car still prevails in the territory due to large distances and non-optimal public transport.[66]

In most European cases, although the provision of public transport is far much more developed than in the Southern Mediterranean countries, only railways and motorways connect the New Towns to their mother cities. However, the connections between the New Towns and other peripheries and even within the New Towns' territories are less efficient: for example, nowadays it is still necessary to take a train back to the capital city (Paris or London) to travel from a New Town to another. The growth of peri-urbanized areas was blocked by such transport problems. Milton Keynes, for instance, has separate residential and industrial grid squares, meaning that zoning was particularly marked. Employment is dispersed across the town in different locations, and population density is extremely low, requiring people to travel long distances between activities. The needs of the motor car have determined alone almost every major element of this plan. To benefit from living in most New Towns, a car is essential.[67]

In existing Egyptian New Towns, like in the city of 6th of October, the absence of an easy transportation system is seen as one of the explanations, together with the high prices of housing, why the residents have not yet migrated to the New Town. Some areas of the town still remain empty, and this illustrates one of the main challenges of the Egyptian new cities: to attract residents and convince them to move to the desert areas.[68] The New Town of Sidi Abdellah is studying the possibility to implement a tramway to and from Algier, while New Towns located in

---

66 Schémas de Cohérence Territoriale, Val Maubuée/Marne-la-Vallée, 2012.

67 Hobson, J., "New Towns, the Modernist Planning Project and Social Justice. The Case of Milton Keynes UK and 6th October Egypt", University College London, 1999.

68 Goell, E. *et al.*, *Sustainable Cities in Egypt. Learning from Experience: Potentials and Preconditions for New Cities in Desert Areas*, The Egyptian Cabinet, Information and Decision Support Center, Center for Future Studies, 2009.

the desert present the most critical situation in terms of transport provision (they are created as self-contained communities), from scratch presenting enormous challenges in terms of development and not less enormous threats for their surrounding environment.

C. *What About Housing and Communities?*

The first aspiration for most of the people is housing, and the provision of this housing is the most common expressed objective of New Towns although most newly developed projects try to create towns integrating all the different functions: housing but also employment, culture, education, sport, leisure, retail, transport. In European urban history New towns were built mainly to provide accommodation for people. Everywhere where New Towns are being built however, it seems that this specific housing or "dormitory suburb" function has ended and that planners are more aware that the expectations from the residents are more than just having a place to live.

As we have explained already in London after the World War II, the creation of the New Towns would lead to the relocation of two million inhabitants. The dwelling units built at that time, however, were designed to answer to the demographic patterns of that period: the creation of large housing estates and scattered pavilions for nuclear families (two parents and two children) mainly from middle-class areas.

With the aging of New Towns, the housing patterns have needed to adapt to new realities:

- The arrival of poorer populations, including from vulnerable groups with larger families
- The aging of population, and the necessity to adapt infrastructure and facilities (lifts or elevators in the housing estates)
- The diversification of housing needs (generalization of people living alone or without children).

These questions however show how much housing was at the heart of the creation of the New Towns. But only the housing function does not go without difficulties: the right identification of future "customers" or inhabitants seem necessary for the planning of housing areas. This is why, among others, the experts from the New Medina project have recommended the creation of an observatory for housing needs in Chrafate/Tangier to adapt the housing offer to the needs.[69] The risk, in the creation of the New Town, is again the one of speculation with

[69] See: New Medina Chrafate, 2nd study.

buildings remaining empty like in the case of 6th of October[70] or New Fayoum.

Housing areas, city of 6th of October Egypt, 2011

Construction of housing in New Fayoum

The question of creating New Towns also generally means the possible relocation of existing villages, the reduction of agricultural land, and the changes of lifestyles for the initial population that will maybe not be able to afford to live in the newly created New Town because, among others, of non-affordable prices. In the New Towns that are created in the desert areas, this question is less important because there are few inhabitants in these zones. In some cases, accommodation is provided to

70 Yousry, S.A., *op. cit.*

the people and families that are expropriated. However, the new homes do not always offer the proximity to services and the possibility to continue the same professional activity (often agriculture or cattle). As a matter of fact, we assist a redefinition of space where rich people that are potentially identified as "future inhabitants" will replace the local population which is relocated sometimes regardless of their existing social, familial and professional connections. It does not prevent in some countries the potential arrival of rural migrants to the newly rich area to counterbalance these disparities.

D. Ecology and Environment

Each city nowadays wants to be "branded" as sustainable. Although New Towns were inspired from the "*garden cities movement*" the question of environment was not a real concern for the decision makers in charge of designing the New Towns some decades ago. The concept that was prior was to have "green areas" and "green towns"...without real consideration for the environmental impact. In particular what has been called "the garden cities movement" promoted the idea "to bring the city into the countryside, and the countryside into the town" but proposed a really scattered urban pattern that would be called today "urban sprawl" and is highly criticized for the ecological footprint. The green utopia was also present in socialist countries like in Eastern Germany with the creation of green garden towns like Eisenhüttenstadt that are representative of the political regime of that time.

New Towns in general had to be green, and the housing areas scattered on the territory. They had to be built on a polycentric pattern. It explains nowadays that the density of New Towns is quite low and some of them lacked a proper town centre for a long time.

Some decades ago, it was difficult to imagine that resources such as oil would end up. Cars were the symbol of progress and therefore New towns and the urban development projects of that time were mainly designed for the use of individual cars, with long distances between the housing areas, shopping areas, and public facilities like schools.

It is only since the first oil crisis in 1973, with the rise of oil prices that questions such as environment, public transport, and quality of air/soil have been taken into account by the planners and decision makers. Nowadays in the European New Towns, the questions of environment, including biodiversity, transport, lighting and reduction of CO_2 emissions have become more important. Each town has a department for environment and each New Town has invested in different areas: In Marne-la-Vallée/Val Maubuée, investments are being made for geothermic production of energy, and a cluster for sustainable construction is being developed with the university.[71] Zoetermeer is developing pilot projects on renewable construction areas etc. It is however generally acknowledged by the promoters of the environment/sustainability that New Towns are representatives of a generalization of urban sprawl and favour "commuters' movements" between the capital cities and the peripheries.

In the New Medina Southern Mediterranean partner towns, and in the different panel discussions, the question of environment has been very much present in the debates. New and innovative technologies are being proposed to build more ecologically friendly towns: wind energy production in Chrafate, renewable energies in Boughezoul, eco-construction in some of the Egyptian New Towns promoted by the HBRC (Housing and Building Research Centre).[72]

In Egypt, a referential for sustainable cities is being discussed. As Pierre Arnaud Barthel also shows in his research on eco-areas,[73] there is a will to build some environmentally friendly areas, to develop pioneering actions for the environment, and to use more environmentally friendly technologies in the areas of waste management, lighting, etc.

This concern nevertheless shows some limits in the implementation. "The concern for environment, and the awareness of the limits of eco-systems have progressed, but they do not lead to common strategies, including in Europe".[74] Indeed, only limited technologies are available and the difficulties for their use in the New Towns' design prevent the real development of more environmental towns.

The reality of their development shows a difficulty to evaluate the risks for safety and for the environmental impact. The questions of public transport and accessibility, but also uses of eco-materials produced locally on the territories, or used by the methods of traditional building often more adapted to climate, are not prioritized. The question of the use of water for the cities of the desert may have some impacts on the desertification of arable land and agricultural production.

Also in comparison to other problems such as informal settlements, urban poverty, absence of sanitation in old medinas, the questions related to ecology and environment can appear as a problem of the "rich" and "wealthy". But good practices are also being found, devel-

71 Cluster Advancity.

72 One of the partners of the New Medina Project.

73 Barthel, P.-A., *op. cit.*

74 Balbo, M. (dir.), *Medinas 2030, Scénarios et stratégies*, Paris, L'Harmattan, 2010, p. 28.

oped and replicated in southern countries and can serve as models for future sustainable urban development.[75]

Eco construction building HBRC

[75] The HBRC proposes pilot eco construction sites. In the panel of experts in New Fayoum Pr. Mario Grosso also made a presentation of eco-construction techniques developed in southern countries. The pending question remains the application at large scale of retrofitting and eco building techniques.

Figure 3. Comparative Table of the different New Towns' contextual background, projects and state of evolution

New Town	Ch'rafate	Sidi Abdellah	New Fayoum
Green belt		●	
Green spaces	●	●	
Proximity of a large city	●	●	
Arable land		●	
University			
Cyberpark/Cluster		●	
Proximity of an industrial site	●		
Charter for architectural quality	●		
Integrative actions for existing villages	●		
Environmental actions/ concerns	●	●	●
Public transport provision		●	
Provision of services/ schools leisure	●	●	●
Zoning	●	●	●
Use of eco materials/ local materials			
Studies for risks such as flooding	●		
Urban marketing	●	●	
Prospective for housing	●	●	
Less than 30% built	●		●
Service ducts		●	

E. The Management and Governance of the New Towns

As it has been described earlier, New Towns are not planned to remain new but rather to become real cities and liveable cities, enjoyable for the inhabitants. Although they are built on the long term, the question of the transition should not be a far horizon that justifies that all decisions are taken arbitrary or quickly. The question of management of the New Towns is becoming increasingly important with the time/step of their development.

The European New Towns are nowadays not only focused on further development or on expanding further, but rather on their management, and the one of their community, territory, environment, and on resources, including the questions of budget (resources/expenditure).

This management is confronted most of the time with an increasing need of local resources, when the public subventions or fiscal resources are becoming scarce. When the transition to a "normal" statute is difficult, it creates some difficulties to maintain the public and green spaces, and the creation of difficult areas, or areas that need to be regenerated. The case of British New Towns is quite self-explanatory. After the New Towns Development Corporations have been wound up, the local authorities did not have enough resources to manage the greeneries, public spaces, museums, etc. This created some pressure to find private funds to continue the town, especially for the further development, or redevelopment of the towns' centres. In the European New Towns, the question of regeneration is indeed a real challenge, as the buildings have been constructed very quickly and the physical infrastructure is ageing simultaneously unlike what happens in older cities.

The difficulty of this challenge is to identify the right level of governance for the future New Town, as it is exemplified in the case of Chrafate. The New Town needs to be managed by a local authority that has the right level of decision-making, and resources, including staff and capacity, to manage the New Town. It also needs to be integrated in a wider territory, with cooperation and identification of competences at the different levels of decision making with the different ministries in charge, region, province, governorship, or Wilaya. However, cooperation leads to the necessity to create other structures like inter-municipalities, with the risk to create extra complexities for the future management of the town.

The question of the involvement of inhabitants in the future design of the town and decision making is also raised in all the discussions and meetings, with all the difficulties and questions that it can create: how to organize it? How to manage contradictions? Are more consultative forms of planning more successful and for whom? All these questions have not disappeared from the discussions in Europe. The difference is however between countries where the choices for the place of living is determined by individual choice, employment, family, transport, costs and other factors[76] and the countries where the power is centralized and it can decide to "send" activity areas and residents to some newly built areas. The creation of New Towns, however, always triggers the question on how to attract the first inhabitants and how to keep them.

Conclusion

There are very few indicators yet on how to identify a "successful town", livable, enjoyable far from the images of ghost towns, towns without souls or dormitory cities. Some of the founders of the French New Towns, would have said: "A New Town is finished when its cemeteries are full".[77] Indeed the very long term is the most important factor of the development of any town, but New Towns also have a specific cycle of development, with a necessity of continuous transformation and adaptation to new challenges, new realities, new priorities and new data. Not only does the success of New Towns come from the inhabitants but only they can "make the town" lively and booming. That being said, the New Medina participants have tried to identify some areas and success factors that seem logical while designing these New Towns.

- The integration, absorption within the territory's patterns and population
- The absence of "copy/paste" models both in terms of construction and planning
- The integration of local know-how and materials within the design
- The adaptation to the context, and evolution of plans
- The global/integrated approach including the economic aspects
- The marketing and definition of image
- The governance
- The capacity building of the local representatives to urban management
- The adaptation of short term planning with long term development
- The integration of the environmental impact

76 Van de Wee, B., Conference European Metropolitan Institute, Brussels, May 2012.

77 In particular P. Delouvrier general delegate for the development of the Paris region, 1961-1969.

Many questions still remain and are pending after all the discussions of the different events.

- What models do New Towns have? Is it the global modern city with skyscrapers? Are they the historic cities? What are the models for small and average developments that do not want to become dormitory cities?
- The question of risks analysis and prospective: are the studies undertaken made to prevent risks or to rationalize already taken decisions? How to cope with the risks of financial setbacks?
- The question of image and identity: are campaigns "soon to be built",[78] like there are done in Europe, sufficient to match the gap between the town to-be-built and the town in construction? How to make representations match?
- Finally is a participatory model of planning a New Town realistic and under what conditions?

To conclude, it is difficult to identify what would make a New Town successful. "There is no recipe for building a New Town",[79] if not long-term development, aspirations and appropriation by the future residents. Maybe the best is to keep in mind that every town has, at one time, been a New Town. Older cities are often represented as a "mille feuille" so as to represent the different layers of urban life: natural and physical infrastructure, the movement of economy, people and transport, spaces and landscape, facilities, architectural styles of buildings, lifestyles, communities, ways of thinking, creativity, production spaces and imaginaries. Integrating this multi layered "mille feuilles" into a Master plan still remains a challenge.

Bibliography

Abaza, M., "Critical Commentary Cairo's downtown imagined Dubaisation or Nostalgia", in *Journal of urban studies*, No. 48, 1075, May 2011.

Ahmed, Y.S., Urban Transformations in New Cities in Egypt, case study 6th of October City, Cairo, 2010.

Balbo, M. (dir.), *Medinas 2030: Scénarios et stratégies*, Paris, L'Harmattan, 2010.

Barthel, P.A., "Premiers quartiers urbains durables dans les pays arabes: enseignement sur une génération spontanée", in *Espaces et sociétés*, No. 147, CAIRN, 2011, p. 99-115.

Barthel, P.A. et S. Monqid, *Le Caire: réinventer la ville*, Paris, Éditions Autrement, 2001.

[78] See communication campaign Apeldoorn 2011-2012.

[79] Panel of experts New Fayoum, Jean Pierre Marchetti.

Battesti, V. et F. Ireton (dir.), *L'Égypte au présent: inventaire d'une société avant révolution*, Paris, Sindbad, 2011.

Boumedine, S.R., "Les villes nouvelles aspects sociaux et Environnementaux", Marseille, 2010.

Chorfi, A., "Les villes nouvelles marocaines", in *Le Maroc en Perspective: Regards croisés*, Les cahiers de l'IAU Idf, No. 154, May 2010.

Dorman, J.W., *The Politics of Neglect: The Egyptian State in Cairo 1974-1998*, Thèse de doctorat, Université de Londres, 2007.

Florin, B., "Des cités nassériennes aux villes nouvelles du désert: la fin du logement social?", in Battesti, V. et F. Ireton (dir.), *L'Égypte au présent*, May 2011.

Gaborit, P., "Reinventing the New Towns Mindset", *Nova Terra*, Delft, September 2005.

Gaborit, P., *New Towns: Image, identity, future perspectives*, Brussels, P.I.E. Peter Lang, 2010.

Goell, E. *et al.*, Sustainable Cities in Egypt. Learning from Experience. Potentials and Preconditions for New Cities in Desert Areas, The Egyptian Cabinet, Information and Decision Support Center, Center for Future Studies, Cairo, 2009.

Hobson, J., *New Towns, the modernist planning project and social justice: the case of Milton Keynes UK and 6th October Egypt*, University College, London, 1999.

Le Tellie, J., F. Debbi et L. Amzil, "La mobilité urbaine dans l'agglomération de Tanger; évolution et perspectives", Plan Bleu, October 2009.

Merlin P., Sudarskis M., *From Garden City to Urban Reconstruction: New Towns in Perspective*, Paris, INTA Press, 1991

New Medina Project, http://www.newmedina.eu.

Report and presentation of Chrafate's panel, project New Medina, November 2010.

Report and presentation of Sidi Abdellah's panel, project New Medina, November 2011.

Report and presentation of Marseille's panel, project New Medina, March 2012.

Report and presentation of New Fayoum's panel, project New Medina, September 2012.

Report of the different studies collected during the project New Medina.

Sims, D., *Understanding Cairo. The Logic of a City Out of control*, Press of the American university in Cairo, 2010.

The European New Towns and Pilot Cities Platform, http://www.pilotcities.eu.

Journal Stratégies internationales spécial Algérie, 4 juillet 2012.

Van de Wee, B., Conference European Metropolitan Institute, Bruxelles, May 2012.

Yousry S.A., *Case Study: Urban Transformations in New Cities in Egypt, 6th of October City*, Cairo, 2010.

A New Town? Where? Why? How?

A Critical and Comparative Approach to New Town Projects in Maghreb

Rachid Sidi BOUMEDINE

Expert and Associate Director at CREAD

Introduction

The topic of new towns in Algeria is one that returns to the agenda seemingly in cycles (which, in turn, puts various potential issues at play).

The ways in which new towns are presented vary little and generally contain two, often-intertwined aspects. One, which is the new town as a cure-all to urban situations (which themselves are presented as inherently negative) and the other, is the new town as project, i.e. its substance (content).

Indeed, in a certain number of cases (for example, in Constantine or Algiers) the proposed new town project is viewed as a harbinger of solutions to the problems seen in the large urban centre they surround (i.e. high population density, lack of amenities or facilities, deteriorated condition of the built environment and transport networks, overcrowding, etc.). They are also touted as vehicles for a new kind of town (and life), as conveyed by the presence of high-level scientific or technological institutions.

I. New Towns: Reducing Congestion

The struggle to reduce traffic congestion is not exclusive to Algeria as it has also been exposed in Morocco and Egypt, specifically in the cities of Cairo and Casablanca.

During the 1970s and 1980s, one of the principal concerns among decision-makers in these different countries was the decongestion of large urban centers (in Cairo, Algiers and Casablanca). These concerns, progressively, opened the way for the formulation new towns. This is to say that, little by little, the stated policy of containing sprawl was substituted by a discourse on land use reallocation, at least in one aspect.[1]

On the other hand, while there is no easy comparison possible between the place and the respective weight of Cairo in Egypt to that of Casablanca, Algiers and Tunis in their respective countries, they nevertheless have been subject to parallel reflections. This was especially the case in the early 1990s, relative to their status as metropolises (and do they share this characteristic?) and to the need to provide them with appropriate amenities and infrastructure.

These developments have been carried out alongside the now ancient and recurring theme of de-congesting the capital cities, which are centers attraction for populations, by means of controlling growth along the peripheries. Seen as sprawling and anarchic, this type of growth is both increasingly important and restricting.

The relatively recent issues facing new towns in the Maghreb must therefore be placed in the context of their dual character as projects of the state, taking into account their two primary axes: the location and the content of the new towns.

In terms of the contents and desired outcomes of these new towns, the choices made in the 1970s and 1980s, affected the development of these towns and have, at the same time, evolved with both local and international contexts as well as with the latest visions and trends in urban, architectural and environmental milieus.

After the developmental period that followed their political independence leading to varying degrees of economic and social changes, the countries of the Maghreb found themselves deep in a world defined by new ultraliberal rules. These rules, having become the new standard, applied to environmental matters: biodiversity and the fight to reduce greenhouse gas emissions.

New Versions of New Towns?

This stated "globalization" has, in effect, introduced new operational concepts for the countries of the South. They are now expected to attain competitiveness and attractiveness through a diligent search sustained by economic efficiencies that are attached to a territorial dimension, privileging (or encouraging) organizations at once economic and spatial

---

1 In a number of cases, this was (and still is) applied to housing or habitat extension zones, which are loosely located near amenities (not more or less so than in the cities, whose urban fabric they complement). In the absence of available land, these zones are used to launch developments with the hope of ameliorating the problems of the large urban centers.

(designated cluster type). They are expected to become centers of excellence or of competitiveness.

The countries of the Maghreb must thrive within the context of globalization as well as within the bounds of the new rules this evolution has set forth. Consequently, it is impossible to treat the topic of new towns in the same terms as in the 1980s or 1990s, as it is now necessary to maintain a critical vigilance and continual revaluation on the issues and points of view facing them.

Indeed, in recent years, whether the development of new towns is given over to a systematized and organized process, however ennobled by the words, *new suburbs*, or whether the creation occurs on green fields, a highly symbolic place is given to the advanced technological discourse in the vocabulary of the new city. In terms of image, this represents a jump in the times (in which one would use and develop these technologies in one's country) and a qualitative jump ahead towards the idealized future city: which explains in part the place given to futuristic and virtual illustrations so beloved by decision-makers.[2]

Whatever might be the reality of a development, it brings forth a strong ideological charge in the form of an Algeria (emerging from a Maghreb) engaged in modernity and the sciences. There is an aspect of magic contained in this idea of the new town: conveying that what is left is only to build sophisticated urban housing zones (ZHUN)[3] in concrete.

Therefore, the concept of the new town covers, if not in its political discourse, then in its very making, an entire series of objects that emerge from the New Town itself. In other words, they are created from scratch on a territory perceived as vacant and given autonomous, if not singular functions. These functions are then associated with adjacent and or existing urban areas. Linking these separate kernels to existing cities allows them to serve as city-dependent satellite towns.

In the case of annexation by a major city, these satellite towns also have a regional and local function: they contribute to a rearrangement and a redistribution of the population and/or activities within the urban region under consideration.

---

2 One cannot limit the use of the image of glass towers and outlandish parks to the copy of (new) Gulf cities: in this framework, not only do we create "territorial marketing" which requires new doctrines of international law, but this shows a capacity to move with the times, and it shows readiness to play the game of globalization; these practices are therefore also a display of modernity.

3 New Urban Housing Areas (Zones d'habitat Urbain Nouvelles).

II. A Few Starting Points

A. Town and Project

For reasons elaborated below, it might seem unnecessary to investigate if a certain project raises new terminologies (or as it were, nomenclature) of the new town concept or not as it is evidently an urban development project.

In effect, from the moment *a veritable new town is revealed, like in all towns, through its very existence, its reality, and its assumed functions*, it is futile to exercise a premature censorship against those who display such a project and/or to question the legitimacy of their viewpoint.

By contrast, one does reserve the right to discuss a project's relevance much in the same way one critiques that of any urban development project: in terms of its feasibility. If the minimum requirements are not fulfilled, it will be, like many other types of developments, a sum of documents to fill the files of urbanist utopias or simply exercises in style.

Sometimes the debate, like that arising on the subject of Boughzoul, in Algeria, concerns not just the opportunities they bring or the programs they contain, but also their location. Those in defense of a project attribute to it all virtues, such as the ability to create a node structuring the Central Region; capable of slowing, if not reversing, the catastrophic migratory flows towards "the saturated north." On the other side, detractors point to the spatial dispersal of (all) new development, its emptiness and its artificial character, or the supposed inability for those in charge to follow through with a development's promises.

But the essential debates continue to focus on the content (the program) of the new city (and also of each new town project) and, to a large extent also on its development.

All these debates are arranged as though the new town is a given, appearing once and for all, and that its contents (the functions which justify such a project) and its physical shell (built structures and infrastructure) would take place on the territory, in the mesh of existing towns (which themselves are constantly changing). In other words, appearing instantly on the land, container and contents conjoined and lapped to each other.

However, a city or town (no matter which) is the observable result, in a given moment, of regular adjustments or even disturbances among its functions. It is comprised of multiple, varied levels of functions and containers (built environment, infrastructures), differentiated by their scale, condition and location.

And, as history also shows us, a city takes root within a national and sometimes international context of activities that give it its place and that influence its expansion or decline.

This brings to bear the error or bias in these aforementioned debates: which is to propose a state zero or blank slate for the new town, i.e. as if it was simply a construction in no greater complexity than materialistic; lacking any other dimensions. But in reality, even if we look at it in its material dimension as first an urban project, rather than as an emerging city, we would still consider it an evolving ensemble of objects, increasingly important and complex.

All this is to say that when discussing new towns we speak of something whose character is in constant flux, resulting in a conception and realization to voluntarily go towards an ever more autonomous ensemble, one which is full of life and capable of functioning and fulfilling its purpose (whether this was initially intended or not); an ensemble capable of inserting itself and adapting to a physical, environmental, as well as a socio-economic context, and therein rooting itself.

Before proceeding to the comparison of new towns in the Maghreb using case studies Chrafat (in Morocco) and Sidi Abdallah (in Algeria), I would like to add a few specific comments to highlight the difference between the new town situation in Europe versus in the Maghreb in terms of experience, but also to touch upon a particular and often insufficiently addressed dimension of how the new town is "creating itself" in the Maghreb context.

B. New Towns: An Agreed Upon Terminology

It is, in fact, notable that often the debate surrounding new towns deals with their intramural aspects, variable in their content (their composition) and their objectives without treating their highly variable external conditions, including their economic, social, and territorial implementation, despite their extreme importance.

The preceding statement[4] makes evident the difficulties of the comparison engendered by the "suitcase-concept" of the new town, the purpose of which is to be able to draw foundational lessons, but which also exists because of the great difficulty of transposing these to a different context, being in the same country.

---

4 Here I am referring to the question that I brought up in a preceding encounter, organized by the Town Ministry concerning "*Social and environmental impacts of new towns: a few preliminary observations*", in Communiqué to the International Forum on New Towns, for sustainable metropolises, Marseille, September 30-October 1, 2010.

As stated above, it seems that what all new town projects have in common is a projection in the future, as they are often presented in a "finished" state, a finality (totality), which seems to be presented as evidence but also as a form of proposed modernity, a combination of a prefigured development of the latest technologies and new forms of economic activity (corporations and urban tissue).

And the experience in Europe, particularly in France, seems largely to illustrate this schema. I'll add that the slow and patient work of realizing these projects, which began at least forty years ago, has allowed for the accumulation of considerable knowledge and has allowed new competencies to emerge.

C. The Context of the Town, Place or Cause of Impact?

From the moment one seeks to draw a profit from this immense experience in which reflections on territorial effects of the placement of a new town have contributed in great part to the enrichment of the reflection on the interactions between urbanization and environment and in the same way have refined the notion of sustainability, one realizes the difficulty comes not from the concept of the new town itself, but from the apprehension surrounding its establishment process.

In fact, experience and lesson seem to be intimately tied to context, which I would globally quantify as the presence (strong, medium or inexistent) or lack of social and economic tissues, infrastructural and technological, that reach or multiply a varies of impacts, positive or negative, of the project, and, I would add, in perpetuity…

Also, an approach to the issue of new towns, through the current Algerian experience, demands not only reflection in terms of the content given to new towns to be indicators of what is to come, but also poses the question of knowing if the context is able to permit the dialectic which is indispensible for the creation of positive impacts.

To summarize, in Europe, one would say that the new town experience has been that of management of cities rather than the launching or creation of new towns, as it has been in developing countries. This difference underscores the issue of context in all its dimensions.

From this point of view, that which considerably differentiates Algerian towns from each other comes as a result not so much of their composition or their formal mission to recompose spaces (economic, technological, or territorial) but in the context of their establishment.

In fact, and according to the case study, the context of a new town can act as a favorable factor as it joins or contains elements favorable to the project (presence of or concentration of human capital, infrastructure, etc.). In contrast, the lack of these elements adds to the scarcity of elemental resources like water or existing handicaps like high unemployment or the under qualifications of local labor.

To refine our analysis, it is necessary to introduce new dimensions in the treatment of this issue affecting new towns: contrary to the work of BTP (buildings and public works) that becomes operational, in terms of finished construction (keys in hand, according to the local expression), after their reception, the new town does not open itself up to the end of a cycle, rather, it takes shape and life progressively, between ten and twenty years, and in fact, is never a finished object, even thereafter.

The new town begins to have large impacts even before it is "finished" being created.

D. The New Town and the City Making Each Other

Indeed, within a regional network dense in services and population, the new town offers much due to its relative mass (low) as well as its contents; a form of re-composition of these networks, since its mass does not decisively influence the regional context nor does it upset its balance. During the phase of new town project development, construction tools and labor are present in this tissue, in the form of hundreds, if not thousands, of sites in parallel with the region.

In the case of projects like Boughzoul, situated in a steppe zone of low population density, and largely under-equipped (even including towns within an 80 km radius), a new town project represents a considerable upheaval due to the mass of jobs it brings, the demands on services it creates, and by the revenues it distributes to workers, subcontractors and craftsmen, etc., in a poor area.[5]

Also, regardless of the composition a new town project to be built, it begins first as an implementation of the construction and public works sector. This sector commonly employs unskilled manpower and the construction of new town projects is long and complex. The cases of El Minya, situated in the middle of the desert, and that of Hassi Messaoud, are similar to that of Boughzoul, even if the latter's influence is not as vast.

That which gives consistency to and impacts a project throughout the years, will be its site characteristics and not its "face" as a technological town or some similar broad concept.

What then justifies a population settling in a new town, giving it its substance? After all, if a city's current location satisfies all needs, there is presumably no reason to relocate it. This kind of discourse occurs

[5] Named as "improvement zones" in the official terminology due to its lack of resources and low-income population.

when new towns are introduced, as if the future new town had to have a strong attractiveness for its future leaders, well before physically existing enough to be considered a town!

E. City and Activities: Stay or Leave?

It is under these conditions that a preliminary question is raised: how does one give living substance to a new town? One could imagine voluntary initiatives, but this must be efficient and credible enough to compensate, for example, an entrepreneur to give up daily business operations for a minimum of three to five years (necessary to attain growth to establish a solidly loyal clientele).

An aspect closely related to the preliminary question (raised above) is the issue of belonging or the insertion of activities into the new town: the specific nature of the city offers each of the economic actors who operates within it (or who is destined to do so) a concentration, or, at least, a certain number of clients (demand) and the supply of services indispensable for it to exercise its activities and achieve profitability.

If the conurbation factor reaches the scale of the city (and the larger the scale, the more one can expect this multiplication of actors useful in one way or another), it acts also upon the regional scale and network of clients and suppliers that exist within it. This underscores the importance for a company, of the context in which it performs, as it is, in fact, a vital tissue to consider.

For this reason it is necessary to take this context into account as an element advantageous (or disadvantageous) to insertion of a proposed establishment: the observation of areas that appear without merit but which nevertheless achieve success in taking off could be illustrative of undetected positive factors which should be accounted for.

Yet we fall silent when confronted with factors resistant to change. Social relations, whether kin or friendship between "candidates for displacement" and their living environment, are the forces that provide the territorially defined feeling of belonging in a group. These aspects of identity are not minor issues to resolve, neither in terms of displacing populations nor in the name of hosting them.

As it appears through successive iterations, the creation of a new town in terms of its active tissue cannot be reduced to achieving a balance between its built environment and the activities and functions it is supposed (or projected) to house, but that this reasoning extends to the entire site (the region) where a new town is established.

In other words, whenever a town is created from scratch, what is needed is to generate the right conditions by which to populate it (if not to furnish it) at the same time that one tries to enliven the city. While it is relatively simple to resolve problems of amenities and infrastructures networks, it is the composition of settlements and activities that is the real issue as well as the primary challenge.[6]

F. New Town on a Virgin Land?

Outside of rare cases or abstract notions of 0.5 inhabitants per km², symbolized by deserts, one concludes upon examination that virgin land is non-existent. There is always some population or settlement making use of the land. Or, in the case of land traveled by nomads, the material occupation is temporary or episodic. We can consider these lands settled (since they are not without their master), and this fact often comes to light when an outside party intends to move there permanently.

There are cases of conflicts, which are unrelentingly renewed and are born out of thoughtless apposition of municipal boundaries, however symbolic they might be, upon rangelands deemed to "belong" to tribes living elsewhere (i.e. in "other communes") by the grace of administrative divisions.

But, what is the meaning of a town in spite of its artifacts of pseudo-egalitarian norms of habitat and amenities? The term pseudo-egalitarian is applicable because the relationship, for example, between an amenity and a certain population size gives optimal service to more concentrated populations and disadvantages those which are dispersed (or less dense); what to say then about the advantages that a strong population density has to access amenities of a higher caliber!

Inequality is even more apparent when it comes to housing, because so-called urban housing is, according to official standards of surface, comfort, and credits relative to it, always (accorded different names) far more vast, better equipped, and more comfortable, than those destined for populations deemed rural. Can one ever assume the pre-existing inhabitants (rural by definition) of the new town project development areas will remain quiet in face of the appointment of their lands to outsiders, considering that these outsiders will have access to housing, services and amenities never once provided to them? The thesis of vacant or semi-vacant land, which underlies the choosing of land for

6 It is due to the sometimes-willful ignorance of this basic principle that the creation of new towns generates considerable commuting. For example, this is seen in Egypt, between Cairo and the new town in question. A similar conclusion was made in the case of Chrafate, located southeast of Tangiers. Living in Tangiers offers plenty of amenities in addition to its agreeableness. Does this fact not risk the rejection of the new town by those who are supposed to move there?

establishing new towns tends to ignore the actors who already inhabit there, those who occupy and make use of land to be developed.[7]

This means it is important to introduce this category of actors when reflecting on the choice of sites and village typologies, and consequently, the nature of activities planned for the town. As the new town represents also a new type of sharing, one has to take into account the rights of these prior inhabitants, and "negotiate" the town with them, or at least the details of the establishment of the new town.

Avoidance of these factors, which might seem minor, often leads to the birth of long-term conflicts that become irresolvable, as they feed from secondary conflicts.

G. Towns: Product of Actors

The issue of actors who make the city take on its importance, especially as it expands to a larger ensemble of towns, and, as well, to the ensemble of activities that encompass the city, is quietly introduced. Since the middle of the 1980s and the debut of the 1990s, it bears mentioning how much the State has withdrawn from public services and productive investments in Algeria and other countries of the Maghreb, as well as how much, in parallel, the private sector (whatever label one has given it, formal or informal) has prospered.

But, as studies have shown, this growth has not obeyed or incorporated into its plans the expectations or necessities of the geographical location of the State. On the contrary, in addition to the emergence of unexpected territorial specializations, which have given another face to the urban framework, one can say that the reshaping of the economic map of the country by its private actors has surprised many.

Given the scale of these changes, sensitive despite the respective weight of the cities and their functions in the late 1970s, and the nature of the changes generated on the network of towns, it is almost impossible to ignore the strength of the entrenched mechanisms and dynamics at the moment of adopting the principles of constructing a new town.

Is it indeed possible to create a new town without the aid or membership of key stakeholders who could be interested, in one way or another, in the creation of a new human settlement? Would it not be preferable, in this case, to opt for a reconfiguration and reshaping of one of the already built towns, knowing that, traditionally in Algeria, this type of

---

7 This issue was rightly raised by panel participants regarding Sidi Abdellah, following Marseille, to bring up the place of peasants who live on the site of Sidi Abdellah (and the environment they have created, the agricultural exploitation and the landscapes) in the process of the confection of the town, and insisted on the ways they could decisively contribute to the "fashioning" of a village integrated into its context.

town, (classed as second-class facility according to authorities)[8] is already in the process of being formed or of prospering in its informality?

As long as the economic and territorial dynamics, which serve as context to the design of a new town and its implementation, are ignored, there can be no effective and sustainable project. This indicates the importance of having an observatory of urban actions as well as the pertinent tools to carry out these observations. Paradoxically, the figures are not lacking, but they are firmly fixed in the categories of analysis produced for the population census of 1966 (and renewed with little subsequent changes), making it impossible to conduct the pertinent urban analyses given the complex aspects of urbanization and the evolution of cities in Algeria.

III. A Dual Core or the Two Functions of the City

And it is not because the current design of new towns, at least in Algeria, is focused on the reception of new technologies that we should lose sight of their other expected roles. That is, as a link in the system of social regulation: the rebalancing of city networks, encouraging a reallocation of populations (wanted or rendered almost mechanical by the specialization of the city), at the regional or national scale!

It is important to keep in mind the central and permanent mission of States, which is regulation, maintenance of social peace and the development, as much as possible, of different territories. Or, if *new technologies, which are currently a core of a part of official new town projects, do not always clash*[9] *with local interests*, above all in rural areas, they however continue to play an effective role in the spatial redistribution of economies and of populations, even though this role is rarely highlighted.

If we introduce this parameter when comparing specific and common problems of the new towns of the southern Mediterranean, *we are led to compare the new towns, not only as advertised projects but also by their territorial position.* This would help to decipher or predict the expected outcomes stemming from their chosen location.

It is necessary to keep in mind that in any event, the establishment responsible for the project manages what is "intramural" while high-

[8] Note that human settlements are classed into urban, semi-urban, and rural. This grade of classifications is oriented to the actions of the State, in terms of the nature and level of services, the type and the size of activities, etc. This is not a play on words; it is enough to compare the configurations proposed by the SNAT and the reality of certain towns' dynamism to be convinced. The contrary is also true.

[9] As it has already been highlighted in different articles in the press, Renault's sole project in Tangier has been confronted with the lack of qualified labor force for the automobile car industry.

ranking authorities make the "outside of the walls" decisions. What this suggests is that the programs assigned to the city (investments, housing and services) as well its infrastructure connections are decided outside of its sphere of influence.

While the focus of this discussion is specifically concerned with new towns (their content, the rules for their implementation, and the associated issues of governance), and, is "intramural," it is important to think about the role new towns play or can play in the context of their implementation, which is never summarized simply through a "procedural choosing of land", as one knows that this context is both used and occupied by stakeholders.

So there exist certain interferences between the levels of power where social regulation (i.e. governance) is exercised, and the level of intervention of planners or developers, the latter of which never appears until one goes beyond the "intramural" context.

In applying this reasoning to new towns near Algiers, specifically in the cases of Bouinan and Sidi Abdallah, both at the heart of the urbanized area of Algiers, the result of their establishment comes back essentially, and definitively, to a cross-regional reorganizing of existing potentialities. This is because the majority of forces that are intended to "furnish" the clusters and the technological parks come from universities and research centers established in the region.

In Boughzoul, a veritable "above ground" town, named so because its scientific activities have nothing to do with the local context, it would at least have a role to play in the territorial network by providing an anchor or pretext for the creation of infrastructural corridors at the national level. East/West and North/South would then cross.

The difference between these two types of towns is not so much in their content than in their location, which confers a different place to the process of implementation and management. In the case of Sidi Abdallah, the infrastructural tissue and the socio-economic support preexist. Consequently, project management missions come back to the essential: that of the internal planning of the town and to the provision of operational space for applicants.

Since the spaces located in the perimeter do not all offer the same opportunities to meet the specific demands of each particular project, there is a risk of fragmenting a new town project into different sub basins.[10] In addition, there are other constraints, posed by the gap be-

10 Kadiri N., Recommendations, Panel of Experts Sidi Abdellah, New Medina Project, November 2011.

tween the levels of development of different areas, and land limitations, all on a territory fragmented by privately property.

However, a large part of the process to locate investments escapes the internal management of the new town and the direction its managers would like to take, in as much as it is a high-level decision. High-level governance and its related issues is then called to generate its own logic; to permanently interfere with the logic of planners who are tasked to plan spatial and functional consistency, and to reduce spatial fragmentation, etc.

In the case of Chrafat, which is similar, the role of the town is not centered on itself and its own activities but on serving as an auxiliary of a much larger project that includes an intercontinental level, a state level, a regional/local level and, finally, an intramural level. The problem is both simpler and more complicated when compared to that of Sidi Abdallah.

Chrafat clearly illustrates this dual problem of regional (and even national and international) scale because it is designed as a complement to a device which contains the city of Tangier, the industrial zones and, specifically, the Renault factories, the ports of Tangier Med and Tétouan and also connection corridors with Europe and with the North/South TGV.

Furthermore, the program of the new town could be run without disruption, as its mission is to open activity areas for outsourcing, and to providing housing areas, as well as urban centers aimed at workers in these industrial zones. By contrast, in Sidi Abdellah, it is activities that give the town its substance; activities that are not programmed by planners or developers but imposed on them.

In other words, it is up to planners in Chrafat to create consistency in programming and phasing of virgin land development. Those in Sidi Abdellah, by contrast, have to face the effects of governance of higher rank. Consequently, they do not master either nature or the timing of choice, and this affects the location process and the ability to manage and mitigate potential impacts. In addition, the town is in an already inhabited and partially structured fabric of hamlets and localities.

This means that the heart of the question of the urban work is not the same knowledge that has to be mobilized in each of these two cases (of towns) at the same time.

This leads us beyond the dual perspective concerning new towns in the Maghreb, to their organization, their content and their structure (intramural) on one side, and, on the other, to the objective role they take through their establishment (extramural). To put the same double glance on the issue of their governance clearly reveals at least two levels, and the corresponding requirements regarding project management.

If one takes into account all these elements, what is essential is the principle of finding a suitable location for a new town and not the implementation (the specific site) of it. And yet, the site may complicate or facilitate the achievement of local objectives or regional distances.

IV. Recommendations and Juridical Context: Practicality and Feasibility of Recommendations

The panels organized at Chrafat and Sidi Abdallah examined a certain number of issues, which I will summarize below into two general categories:

- The first concerns the design of the object "new town": its functions, its internal organization and its location, the concepts put into place, and the project's sustainability, etc.
- The second concerns the general conditions of the project implementation and its forms of development: land management, marketing, the project management, etc.

Nevertheless, as the reason for these panels is to make recommendations, to, firstly, those who carry out project management, it seems essential and rational, to trace the limits of the practicability and feasibility of these panel recommendations for each operator, as the laws of two countries, even if relatively similar at first glance, do not cover the same technical or operational traditions.

These recommendations covered four central themes:

- The description of the object "town" as a project that crystallizes intentions: this means describing the two projects from the point of view of authorities that have decided to develop, using the definitions and the contents they apply.
- The territorial context: this is to take into account the effects of the project's location more than the site of development, and at the same time as an expression of this desire, such as the objective interactions with the context, and the implications on the formal project.
- The legal framework of the action, including the status of actors of the Establishment in charge of the project: this is not to imply that the modes and margins of institutions in different Maghreb countries are therefore comparable, as land transactions, financing methods, and degrees of autonomy of the Establishment are not identical nor comparable.
- The issues of project management (implementation) and procedure: these are the issues tied to technical procedures (consistencies, rou-

tines) with respect to the object, and the issues related to project management by the city (installations, partners, attractiveness, etc.).

The actions of public institutions tasked with new towns in Algeria, for example, are in fact framed by essentially three texts:[11] the law which determines the conditions for the creation of new towns and orders, for each of them, of executive decrees establishing the new town and, finally, a decree establishing the mission, organization and operating procedures of the organization (in charge of development and management) of the new town. Below are references of texts related, for example, to Boughzoul. Included in the appendix is the list of major texts which concern land-use planning.

It is indeed clear that project management has devolved to agencies specifically responsible for each new town (or the ensemble of towns like in the case of Morocco). Three aspects need to be clarified because they decisively determine the margin of manpower and the effectiveness of the organization:

- The mission of the organization and the delegations of transfer in its favour, the powers of public authorities (regulatory or urban policy planning, for example).
- The terms and the conditions of exercise of the Project Manager's Delegate (MOD) by which the State confers powers to the body.
- The objects and the conditions for the exercise of the concept of Public Utility.

In the example of Sidi Abdallah, the difficulties tied to this multiplicity of actors, and also with the varying levels of power and different missions, have been highlighted by the of mobility experts' report. Cited below is one part:

> The seminar (of experts NDR) also found a scattering of powers and responsibilities between different actors and policy makers. The team of the new town seems to be essentially devoted to the technical execution of the operation. Marketing to companies seems to spring from outside agencies. Financial and strategic decisions appear negotiated with the government on an annual budget and finally the overall strategy seems to depend directly on options chosen for resolving the problems of Greater Algiers. This is an acceptable or fertile phase of the project. However it is heavy or sluggish when it comes to taking action. The new town is too fragile to depend on piloting 'outside the walls'. It must rely on itself and deal only with itself.[12]

11 Executive decree number 06-232 of the 8 Joumada Ethania 1427, corresponding to July 4, 2006. Public declaration of public utility published the operation relative to the realization of certain amenities and infrastructure of the new town of Boughezoul.

12 Appearing in "*programme CIUDAD – Projet New Medina expertise de la ville nouvelle de Sidi-Abdellah*" mission à Alger les 21/22/23 novembre 2011 "Préconisa-

Also, the function of these bodies was reduced to project management as a Design Office responsible for executing and monitoring plans without influence over the design or the program of the project, and if they are only allowed, at best, to make TDF a destination of guardianship, the recommendations would be, in their turn, reduced to exercises in style.

However, we know that these organizations have important functions in terms of delegating a project's development, promoting real estate development, management and servicing, without mentioning partnership operations and investments: under these conditions, determining what the law allows can permit fine tuning and adaptation to the recommendations in various scenarios and also allows one to generalize that which is common.

It should be noted that this legal framework is recent, even though the concept of the new town dates to nearly forty years: this signifies that the authorities have passed, in the last decades, from general formulations to a refinement of concepts, tools and types of intervention. This time has not passed without consequences, for the content of such projects as can be seen in the brief history that follows.

V. Some Conclusions

The principal barriers to the success of the new town project (i.e. a real town in a state of opening up and functioning) does not reside in our point of view, in the difficulties of construction of building and infrastructures; that, we know how to do, and is at the limit of paying a price, if one calls upon outside expertise. But what is difficult to achieve, is the creation of a town as a living tissue (either autonomous or grafted) and to ensure that the graft takes hold and contributes to the localization of regions and sectors that serve the city.

It is then convenient that this is not an affair in concrete, but made of economic and social actors, who give value, vitality and substance to towns: consequently, the new town project is primarily a consensus on a social and economic level.

However, perhaps because they do not look for this dimension of the town as a complex tissue graft, resulting in multiple collaborations, Maghreb analysts, in their review of examples of new towns of the Paris region, in Holland, etc. do not seem to draw the consequences of that which, in addition to know-how and technical mastery of these countries, these projects are, in all cases, located within proximity of large

---

tions de marketing opérationnel et stratégique pour la ville nouvelle de Sidi Abdellah".

metropolises, or in areas of high density of urban tissue and activities, they stand to achieve success.

One can easily understand that, in this case, the relocation of operations towards the new site brings material advantages to the entrepreneur (much larger spaces, better conceived and equipped buildings), on top of their financial incentives, which gives life and substance to the new town, in the role of dispositive is an advantage to investors.

What can we then learn from the example of new town in Europe, even though they are not comparable: we have insisted on the fact that the context of implantation is largely more favourable under conditions which prevail in most of the cases in the Maghreb.

By contrast, it seems to me, as the interventions and discussions of the Panel of Marseille have shown, that there are important lessons which remain to be learned from currently observable developments, notably in France; I will touch on at least three, among others, not because they are the most significant, but because they offer us different tracks for reflection.

A. New Towns as Poles in Greater Paris

New towns in the Paris region, including Cergy-Pontoise, Evry, St Quentin en Yvelines, Marne-la-Vallée, were launched in the 1960s at the instigation of de Gaulle and placed under the management of Paul Delouvrier.

They had already contributed, besides the spatial reordering required by the authorities,[13] to give weight to the metropolitan area facing the development of powerful European poles such as the Ruhr and the establishment of an axis from Rotterdam to Milan via Germany, thus marginalizing the Parisian pole.

Forty years after beginning this scheme, which includes the valley of the Seine to Le Havre, as it was envisaged in the studies conducted by OREAM, it remains valid and has adapted to its new context, resting upon on a new project, the Grand Paris, which is based on the new poles of scientific excellence comprised largely by the new towns.

We are therefore in the presence, with the long temporal dimension of planning, of a project that has created a spatial reorganization according to a diagram that has held up, and which shows us the importance of this long dimension as our article focuses on short-term issues, and the

[13] Charles de Gaulle would have said after his arrival while flying over the Parisian suburbia helicopter, to the planner Paul Delouvrier "Please put back some order in this mess".

recent character of the Maghreb show long-term prospective and diagrams of National Planning.

B. Greenery, Landscapes, Farmers and the Environment

Another element, emerging from the presentations and discussions among local actors, to the benefit of all, is on the issue of the formal opposition of occupation.

The merit of these interventions is to bring up the question of a city's "absorption" of rural areas and the possible action in rural areas, which occupy the spaces around the city and represent a different issue in the integration of the space without destroying vegetated or aquatic elements in the city as well as the creation of landscape values. Skills and the use of farmers in this transfer permits thinking on the issue in terms of harmonious elements and actors instead of opposition and incompatibility.

This is even more relevant than the concepts of sustainable city or the green city, as they also materialize with the assistance of local actors, and to the benefit of all, by reducing potential conflicts of occupation.

C. The Economic and Social Actors, Governance, the Ways of Doing

Let us support this issue of the contribution of actors by emphasizing the fact that, as the experience of new cities in France progresses, the place and the role of consultation and negotiation in the conduct of complex projects has emerged as increasingly critical and essential.

Even if the project is still overwhelmingly a State project or of the State, dialogue begins at conception before continuing in the steps related to land, modulation and adjustment projects, avoiding wasted time and money and conflicts, and toward more acceptance and thus cooperation from actors who, ultimately, contribute, even by their opposition to one way or another, to the city.

This teaches us that we should not reduce the management and conduct of projects to technical processes, nor make fashionable slogans, but that we should introduce the philosophy and methods of negotiation as basic elements in the carrying out of projects.

Bibliography

Boumedine, R.S., "Social and Environmental Impacts of New Towns. A Few Preliminary Observations", in Communiqué to the International Forum on New Towns for Sustainable Metropolises, Marseille, September 30-October 1, 2010.

Executive decree number 06-232 du 8 Joumada Ethania 1427, corresponding to July 4, 2006.

Kadiri, N., Recommendations, Panel of Experts Sidi Abdellah, New Medina Project, November 2011.

Programme CIUDAD – Projet New Medina expertise de la ville nouvelle de Sidi-Abdellah, Mission à Alger les 21/22/23 novembre 2011, "Préconisations de marketing opérationnel et stratégique pour la ville nouvelle de Sidi-Abdellah".

Text related to urban planning

Algeria

Loi No. 90-29 du 1er décembre 1990, modifiée et complétée, relative à l'aménagement et l'urbanisme.

Loi No. 90-30 du 1er décembre 1990 portant loi domaniale; Vu la loi No. 91-11 du 27 avril 1991, complétée, fixant les règles relatives à l'expropriation pour cause d'utilité publique.

Loi No. 98-04 du 20 Safar 1419 correspondant au 15 juin 1998 relative à la protection du patrimoine culturel.

Loi No. 2001-20 du 27 Ramadhan 1422 correspondant au 12 décembre 2001 relative à l'aménagement et au développement durable du territoire.

Loi No. 2002-02 du 22 Dhou El Kaada 1422 correspondant au 5 février 2002 relative à la protection et à la valorisation du littoral.

Loi No. 2002-08 du 25 Safar 1423 correspondant au 8 mai 2002 relative aux conditions de création des villes nouvelles et de leur aménagement.

Loi No. 2003-10 du 19 Joumada El Oula 1424 correspondant au 19 juillet 2003 relative à la protection de l'environnement dans le cadre du développement durable.

Loi No. 04-20 du 13 Dhou El Kaada 1425 correspondant au 25 décembre 2004 relative à la prévention des risques majeurs et à la gestion des catastrophes dans le cadre du développement durable, p. 13.

Loi No. 2006-06 du 21 Moharram 1427 correspondant au 20 février 2006 portant loi d'orientation de la ville, p. 14.

Vu le décret exécutif No. 04-97 du 11 Safar 1425 correspondant au 1er avril 2004 portant création de la ville nouvelle de Boughzoul.

Décret exécutif No. 06-304 du 17 Chaâbane 1427 correspondant au 10 Septembre 2006 fixant les missions, l'organisation et les modalités de fonctionnement de l'organisme de la ville nouvelle de Boughzoul.

Décret exécutif No. 06-232 du 8 Joumada Ethania 1427 correspondant au 4 juillet 2006 portant déclaration d'utilité publique l'opération relative à la réalisation de certains ouvrages, équipements, et infrastructures de la ville nouvelle de Boughzoul.

Moroccan studies

Agence Urbaine de Tanger, 2004. Planification et Grands Projets Urbains: Quelle Vision Pour la Ville de Tanger.

Ministère de l'Habitat et l'Urbanisme et de Développement Spatial: Projet de loi relatif aux Villes Nouvelles: Septembre 2009.

Ministère Chargée de l'Habitat et de l'Urbanisme, 2005. Rapport Général de la Première Rencontre des Agences Urbaines Tenue à Fès les 24 et 25 Novembre 2005 sous le Thème: Projet Territorial et Développement.

Ministère Chargée de l'Habitat et de l'Urbanisme (STE AL OMRANE). Etude de Faisabilité du Projet du Nouveau Pôle Urbain JOUAMAA.

Société Al Omrane Al Boughaz Etude de Faisabilité de la Ville Nouvelle Melloussa Tanger 2006.

Legal framework

Loi No. 7-81 relative à l'expropriation pour cause d'utilité publique et à l'occupation temporaire, promulguée par le dahir No. 1-81-254 du 11 rejebn 1402 (6 mai 1 loi No. 1-76-583 du 5 chaoual 1396 (30 septembre 1976) relatif à l'organisation communale. 982).

Loi No. 25-90 relative aux lotissements, groupes d'habitations et morcellements promulguée par le dahir No. 1-92-7 du 15 hija 1412 (17 Juin 1992) (B.O. 15 juillet 1992).

Dahir No. 1.93.51 du 10 septembre 1993 portant création des agences urbaines (JO. No. 4220 du 15/09/1993).

Loi No. 78-00 relative à l'organisation des collectivités locales (JO. No. 5058 du 21 /11/2002).

Décret No. 2-92-832 du 14 octobre 1993 pris en application de la loi 12-90 relative à l'urbanisme (JO. No. 4225 du 20 octobre 1993).

Dahir No. 1-93-51 du 10 septembre 1993 en application de la loi portant création des agences urbaines (JO. No. 4220 du 15 septembre 1993).

Egypt

Law No. 119 for the year 2008 promulgating the building law. *The official gazette* – No. 19 Bis (A), on 11/5/2008.

A Traveling Concept

The New Town Ideal from Howard's Garden City to Today's Ecocity

Deni RUGGERI

PhD Assistant Professor of Landscape Architecture, University of Oregon, USA

I. Setting the Stage: New Towns Around the World

New towns have been built in different continents, political and economic systems, cultures, and climates. Many of them have become important economic and cultural centers in their countries. Almere in the Netherlands, Milton Keynes in the UK, Marne-la-Vallée in France, Reston and Irvine in the USA have developed strong identities, attracted successful businesses, universities and diverse populations. Some have outgrown their original plans, while others have struggled to implement all of their features. This paper seeks to map out the evolution of the New Town ideal over its bicentennial history and reveals that despite sharing a set of core principles, there exist a diversity of nuanced approaches and adaptations of the original visions to fit the idiosyncratic needs of particular times, places, and cultures. Is it possible to speak of common trends in the evolution of the New Town model? And what may be the risks and opportunities connected with the adoption of the model that has travelled to us to the unprecedented challenges of today's world?

New towns continue to be built in some countries. The Chinese government, for instance, has been engaged in the creation of Eco-cities, designed with energy efficiency, climate change and habitat conservation in mind. In a recent article in the Guardian, Deputy British Prime Minister Nick Clegg called for a new round of New Town plans as a solution to the housing crisis in the country.[1] Elsewhere in the world, more traditional New Towns continue to emerge, which place a growing emphasis on sustainability. The New Medina Project managed by the European New Town Platform, is a recent program aimed at establishing collaborations between European and North African New Towns. The final goal is to promote sustainable development practices,[2] building upon the experience of older New Towns in order to lower energy consumption, contain maintenance costs, increase densities, and adapt to changes in taste, socio-economic makeup and identity.[3]

The paper employs Edward Said's "Traveling Theory" (1983) to illustrate how the ideas and designs of New Towns have changed and evolved overtime. It begins by outlining a working definition of New Towns based on the characteristics that are shared by the majority of the world's New Town. It then discusses the "traveling" and evolving nature of the New Town ideals using case studies that best illustrate the evolving nature of this planning paradigm. The paper also seeks to answer a few key questions that relate to the evolution of the New Town Ideal: How has the concept generated, evolved and sustained itself overtime? What have been the most obvious changes and adaptations to the original ideas? It concludes with a discussion of similarities and idiosyncrasies across the world's many New Towns, and highlights potential challenges to the work of designers and planners engaged in shaping the new communities of the future.

An Evolving New Town Definition

Scholars in the planning and design fields have highlighted the multifaceted and evolving nature of New Towns[4] and have been unable to find consensus *vis-à-vis* a definition.[5] In the absence of such a univocal definition, a "working" definition of New Town may still emerge from

---

1 Guardian, "Top Tips: Garden Cities and New Towns", December 8th, 2012, http://m.guardian.co.uk//housing-network/2012/dec/08/top-tips-garden-cities-and-new-towns.

2 Stewart, D.J., "Cities in the Desert. The Egyptian New-Town Program", in *Annals of the Association of American Geographers*, 86:3, 1996, pp. 459-480.

3 Ruggeri, D., *Place-Identity, Attachment and Community Attitudes in Suburbia. The Irvine Ranch Case*, Dissertation, Berkeley, CA, University of California, 2009.

4 According to the INTI, a New Town research institute situated in Almere, Netherlands, there are over one thousand cities around the world who call themselves New Town (http://www.inti.net).

5 Forsyth, A. and K. Crewe, "A Typology of Comprehensive Designed Communities Since the Second World War", in *Landscape Journal*, 28:1, 2009, pp. 56-78; Merlin, P., *New Towns, Regional Planning and Development*, London, Methuen, 1971; Gaborit, P., *European New Towns. Image, Identities, Future Perspective*, Brussels, PIE Peter Lang, 2010.

the analysis of a series of shared characteristics, which may be divided into "core" and "contextual" (table 1). In short, rather than trying to find a one-size-fits-all definition, this paper suggests that we should think of New Towns as a dynamic *gestalt* of "core" and "contextual" elements, and consider the New Town ideal as being a fluid, ever changing concept that is reinterpreted and refined at each lag of its "travel", a concept that eventually becomes crystallized in its built environments.

Table 1. The author's working definition of New Town

Core	Contextual
Self-containment Maximum size and scale (from 32,000 to one million) Master planning and coordination; Regional planning orientation; Emphasis on open space and leisure Balance of residents and jobs Built according to a specific time frame	Reliance on mass transit or high traffic road infrastructure Complete or partial limits to automobile access Mix of uses (housing, commercial, public amenities) Imageability Public interest and policy Coordinated management Technological innovation Presence of universities and other higher education institutions Emphasis on leisure and recreation.

In trying to conceptualize such a New Town definition, I was introduced to Edward Said's *Traveling Theory* (1983). Said, professor of English and comparative literature at Columbia University and founder of the field of post-colonial studies, has crafted a new theory with regard to the creation and re-definition of ideas and discourses as they travel across cultures and political systems.[6] The theory suggests that ideas travel according to four stages: a "point of origin", i.e. a set of circumstances by which an idea is born; a "distance transferred" that accounts for the obstacles an idea has to face before it is accepted; a third stage during which certain conditions allow the "acceptance of rejection of the idea", and a final moment of acceptance and institutionalization. Said's ultimate message is that theories are never complete, and that they cannot be fully explained by a model, or dictionary definition. Rather,

6 Duvall, R. and L. Varadarajan, "Traveling in Paradox. Edward Said and Critical International Relations", in *Millennium – Journal of International Studies*, 36:1, 2007, pp. 83-99.

they should "grasped in the place and time out of which it emerges as part of that time working in and for it, responding to it".[7]

Said's theoretical framework is well suited to the discussion on the roots of the New Town ideal. He suggests that at each stage of the process, it is through the representation of an idea and its institutionalization that the original theory or discourse evolves.

Through representation, theories take on a life of their own; planning diagrams, master plans, and geographical maps become as tangible as the physical landscape[8] and once realized, they become institutionalized in the built landscape of streets, housing, and nature of our cities. Said's four development stages of a travelling theory can help us illustrate how the following milestones in the evolution of the New Town ideal may have emerged:

- A point of origin: Ebenezer Howard's Garden city model;
- A distance transferred: Letchworth and Radburn's garden cities;
- Conditions for acceptance and rejection. Post-World War II New Town planning in Europe and US-Harlow, UK and Irvine, CA;
- A new position in time and place. Tianjin Eco-city and the "eco-smart" Sidi Abdellah New Town.

Table 2, presented below, seeks to map the evolution of the New Town paradigm and its physical manifestations overtime. It shows that while there are many consistent elements, we can point to an increase in the richness and complexity of planning and design features, and an increasing reliance on technological innovation as a source of long-term resilience.

7 Said, E.W., *"Traveling Theory", the World, the Text, and the Critic*, Cambridge, Harvard University, 1983.

8 Cosgrove, D., *Social Formation and Symbolic Landscape*, Madison, Wisconsin, University of Wisconsin Press, 1983.

Table 2. Comparative analysis of New Town distinctive elements

New Towns / Characteristics	Letchworth	Radburn	Harlow	Irvine	Sidi Abdellah	Sino Singapore E-C
Population						
25,000 to 200,000	●	●	●	-	-	-
200,000 to 500,000	-	-	-	●	●	●
500,000 +	-	-	-	-	-	◐
Policy						
Polycentric	-	-	-	●	●	●
Monocentric	●	●	●	-	-	-
Proximity to large metro areas	●	●	-	◐	●	-
Greenfield development	●	●	●	●	●	-
University Town	-	-	-	●	●	-
Company Town	-	-	-	-	-	-
Self Containment						
Economic - Live and Work	●	●	●	●	●	●
Economic - Link to Industry	◐	-	●	●	●	●
Education - University/Educational Institution	-	-	-	●	●	-
Amenities - Retail	-	◐	●	●	-	●
Amenities - Sports and Recreation	-	-	●	●	●	●
Amenities - Regional Parks	-	-	-	●	-	●
Amenities - Urban Scale Open Space	-	●	●	●	-	●
Amenities - Neighbourhood Parks	-	-	●	●	●	●
Amenities - Community Gardens	-	-	●	◐	-	-
Amenities - Agricultural Lands	●	-	●	●	◐	◐
Social - Groups, Clubs, Asssociations	●	●	●	●	-	-
Social - Entertainment / Music, Theater, Sports	-	-	●	●	-	-
Diversity of Housing						
Mix of housing types	●	●	●	●	●	●
Affordable Housing	●	●	●	◐	●	◐
Accessibility						
Internal Public Transportation	-	-	-	●	◐	●
Auto Oriented	-	-	-	●	●	-
Pedestrian Oriented	◐	●	●	-	-	●
Linkages to Natural Areas	-	◐	●	●	-	-
Greenbelt Surrounding the Development	●	◐	●	●	●	●
Bus Access to Surrounding Communities	-	-	●	◐	-	-
Train Access to Surrounding Communities	●	●	●	◐	◐	●
Automobile Access to Surrounding Communities	-	●	●	●	●	●
Pedestrian Access to Surrounding Communities	-	●	●	●	-	-
Supervision of Construction						
Public (Government)	-	●	-	-	●	●
Cooperative	-	●	-	-	-	-
Private	●	-	-	●	-	-
Partnership Public/Private	-	-	-	◐	-	-
Self Generated	◐	-	-	-	-	-
Linked To Technological Innovation						
Energy	-	-	-	-	-	●
Media	-	-	-	◐	-	-
Biotech	-	-	-	●	-	●
Internet	-	-	-	-	-	-
heavy industry	-	-	-	-	-	-
transportation	-	-	-	-	-	-
manufacturing	-	-	-	◐	-	●
other types of industry	-	-	-	-	-	●

● Applicable ◐ Partially App — Not App

Illustration by author

II. The Traveling New Town Ideal

The four milestones described earlier may be illustrated by the planning of case studies from around the world, which were the object of this research: Letchworth, UK; Radburn, NJ; Harlow, UK; Irvine, CA; the Sino-Singapore Eco-city in Tianjin, China; and Sidi Abdellah, Algeria. Beginning in 2010, the author and a small team of graduate students engaged in a yearlong extensive bibliographical and web-based research effort aimed at illustrating the evolution and adaptation of the

New Town ideal overtime. Access to web-based documentation was necessary to fill the many gaps in the New Towns literature, which tends to be thin and not very comprehensive. In particular, the discussion of the contemporary New Town examples such as Sidi Adbellah in Algeria and Tianjin's Sino-Singapore Eco-city in China relied heavily on first hand sources, marketing publications and other miscellaneous advertising material. The data was coded in terms to each New Town's adherence to characteristics and features elements highlighted in the literature as essential to most New Towns worldwide. Adherence or departure from these theoretical models was represented graphically in table 2, which will inform our discussion of similarities and differences in the application of the New Town concept in part III of this paper.

A. Point of Origin

The idea of creating self-contained satellite cities as a solution to housing shortages in fast growing industrial cities of the 19th century can be traced back to the work of British theorist Ebenezer Howard.[9] Howard's own city of London, which since the industrial revolution had become an unliveable mix of housing, industry, and commerce, offered him the point of departure for an investigation of a new and more livable urban form. In his 1902 book *Garden cities of To-morrow*, Howard laid out a model for a regional network of garden cities, self-contained, walkable, and green communities linked by the railroad (figure 1).

Howard accompanied his writings with an iconic diagram showing a new urban structure where urban functions were organized according to a concentric model centered around a central park surrounded by retail and commercial zones, followed by concentric rings of medium to high density residential neighborhoods to accommodate a projected maximum population of 32,000. At the outer edge were industries whose goal was to create enough jobs to ensure the city's long-term resilience. A greenbelt, dedicated to farming, recreation and natural habitat was to be created as a barrier to sprawl and suburbanization, but also as a separation between garden cities. Howard also envisioned a new system of land tenure, by which land would be held in common. By rejecting private ownership, Howard wanted to resolve once and for all the problem of real estate speculators, which had contributed to the degradation of urban living conditions in many European cities. By leasing the land to residents, commerce and industry, the garden city founder envisioned that an average 4% return could be achieved. This was to be invested in

9 Howard, E., *Garden Cities of To-Morrow. A Peaceful Path to Real Reform*, Swan Sonnenschein & Co. Ltd., 1902; Doglio, C., *La Città Giardino*, Roma, Gangemi, 1985; Forsyth, A. and K. Crewe, "A Typology of Comprehensive Designed Communities Since the Second World War", in *Landscape Journal*, 28:1, 2009, pp. 56-78.

future construction, purchase of land as public open space, provision of public services, and as safety funds to be used during harsh economic times.

Figure 1. Ebenezer Howard's Garden City Model

Courtesy: discoveringurbanism/blogspot.com

Letchworth

The first application of Howard's model was Letchworth, nicknamed "the first garden city". It featured livable, walkable residential neighborhoods, a thriving industrial base; a predetermined maximum size of 32,000 on 5,300 acres of undeveloped land, half of which were devoted to a rural greenbelt and held in trust for the community.[10] Its physical form was inspired by the medieval tradition of city building promoted in the writings of Camillo Sitte.[11] In Letchworth, Sitte's picturesque sequence of public spaces carved out of a dense urban fabric had to be reinterpreted and adapted to the low densities of the suburbs. The plan featured a variety of housing typologies, including single family, terraced houses and apartment blocks organized into mixed neighborhoods within walking distance to a commercial center.

10 Osborn, F.J. and A. Whittick, *New Towns. Their Origins, Achievements and Progress*, London, L. Hill, 1977.

11 Sitte, C., *City Planning According to Artistic Principles*, New York, Rizzoli, 1986.

While its original intent was to provide sufficient economic opportunities and jobs, for the first 10 years of its life due Letchworth struggled to attract viable employment opportunities due to its distance from London.[12] Despite initial difficulties, the garden city did eventually develop a strong mix of industry that has weathered war and economic fluctuations. By 1933 there were 150 shops and 60 industrial firms, by the late 1970s the number had risen to 272 shops and 173 manufacturing establishments. This industrial expansion was accompanied by the construction of public housing estates to the north and south of the original town center, and modern suburban homes filling in much of the open space that was originally set aside for the greenbelt.

As the first attempt at the reification of Ebenezer Howard's garden city diagram, Letchworth's founding principles contributed to the institutionalization of the garden city as a legitimate model for urban development. As Said suggests, Letchworth was the original representation of the garden city model, and as such it was an integral part of the theoretical baggage that was exported, traveled and was re-interpreted across cultures.

B. A Distance Transferred: from Stein and Wright's Garden City of Radburn and the post-World War II New Towns of England, France, Sweden and the US

American planners Clarence Perry, Clarence Stein and Henry Wright were among those whose work was deeply inspired by Howard's model community. They saw in the British garden city a model for rational planning and embraced it as a solution to the declining urban conditions of the American city.[13] The walkable neighborhoods, mix of housing, imageable landscapes and balance between urbanity and nature struck those who visited the British garden city upon its completion. These forms inspired the work of many planners, including Clarence Perry, who based his 1929 Neighborhood Unit scheme on Howard's first garden city, thus extending its reach all the way to North America (figure 2).[14]

12 Osborn, F.J. and A. Whittick, *op. cit.*

13 Birch, E.L., *Radburn and the American Planning Movement. The Persistence of an Idea*, New York, Taylor & Francis, 1980.

14 Brody, J.S., *Constructing Professional Knowledge. The Neighborhood Unit Concept in the Community Builders' Handbook*, Doctoral dissertation, University of Illinois, 2009.

Figure 2. Clarence Perry's 1929 Neighbourhood Unit

urbanlandscapes.info

Stein and Wright – both members of the powerful Regional Plan Association of America (or RPPA) – were instrumental in the adaptation of Howard's ideas to America's unique cultural, socio-economic and political context.[15] Radburn, NJ (figure 3) was the first, comprehensive attempt to implement Howard's city diagram on US soil. Stein was particularly clever at adjusting the Garden City theories to the dominance of the automobile and the mechanization and standardization of residential housing construction in the US through the introduction of two planning expedients: the modernist "superblock" and the cul-de-sac. Both were employed to limit through traffic within neighborhoods, thus adding to the convenience and safety of their residents. Also innovative was Stein and Wright's experimentation with modern, prefabricated housing, influenced by the work of their colleague Catherine Bauer's research on affordable housing.[16]

The emphasis on efficiency, lower construction costs and traffic management, embodied by the use of the superblock and standardization in were the most evident embodiments of the influence on American

---

15 Stein, C.S., "Toward New Towns for America", in *The Town Planning Review*, 20:3, 1949, pp. 203-282.

16 Birch, E.L., *op. cit.*

Garden Cities by Le Corbusier's "city for a million people".[17] Stein's garden cities accepted the primacy of the automobile, but introduced an integrated system of grade-separated crossings allowing uninterrupted paths and bike trails. A modernist element consisted the strict zoning, which had been created to protect the integrity of Radburn's plan. A particularly evident element of discontinuity between UK and US revolved around the preservation of agricultural lands to form green-belts. In traveling to North America, the Garden City model lost its greenbelt component, which would have surrounded the community, curbed sprawl and offered residents easy access to natural areas. A look at today's Radburn shows a place engulfed in a sea of suburban tracts.

Figure 3. Master Plan for Radburn Garden City by Stein and Wright, 1929

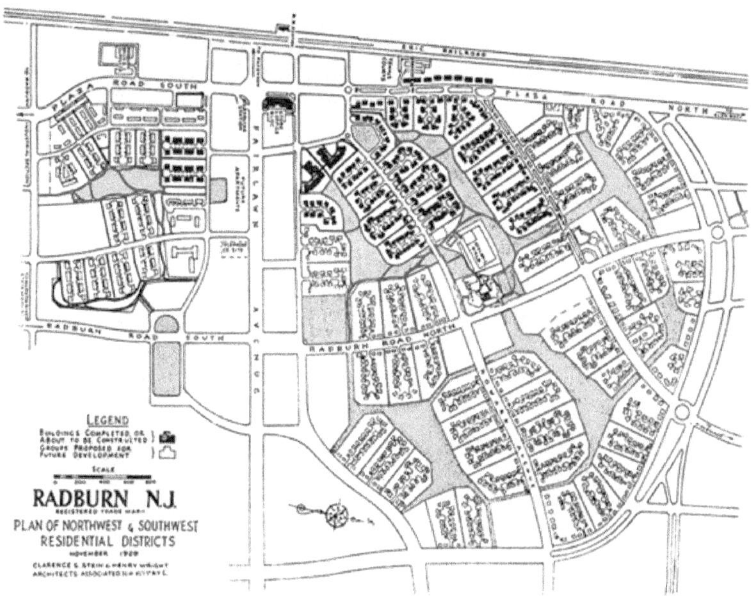

Courtesy: www.fradkinmcalpin.com

The marriage of Howard's Garden City with the Modernist response to the automobile and the industrialization of city building processes were instrumental in the emergence of the New Town ideal, and its diffusion in both the European and American contexts. In the US, the

---

17 Corbusier, L., *The City of Tomorrow and Its Planning*, London, Dover Publications, 1987.

New Town model found institutionalization in the professional figure of the community builder, in charge of the implementation of large scale, comprehensive communities.[18] In the UK and elsewhere in Europe, such institutionalization occurred through the rise of government-sponsored development corporations, in charge of planning and implementing New Towns in response to the unprecedented urban growth that followed the end of World War II.

C. Set of Conditions for Acceptance and Rejection

The post war period witnessed the construction of numerous New Towns in response to housing shortages and unsanitary living conditions of metropolitan areas worldwide. The adaptation of the garden city model and the Modern Movement to the variety of economic, political and geographical contexts where New Towns were built resulted in a few, idiosyncratic types. While in the UK, Sweden, the Netherlands and France national legislation was instrumental in promoting New Towns, other countries preferred to follow a more *ad-hoc* process, while others let the private sectors respond to the challenge as in the case of the United States, Finland and Italy.[19]

In the UK the New Town act (1946) was the result of a governmental effort aimed at the construction of new housing for the displaced residents of urban slums. Most British New Towns had a population between 20,000 and 60,000 people and were located through careful regional planning 40 km outside of London and 20 km from other major cities, thus giving them economic, social and cultural independency from the historic city.[20] In Sweden, New Towns were dependent from central cities, to which they were linked by an extensive subway system. New Town planning in France followed yet another pattern dictated by their organization along a high-speed transportation corridor, and offering mainly housing for the Parisian working classes.

Post-World War II New Towns were built in the agricultural countryside, offering residents living standards and unprecedented access to a wealth of passive recreation opportunities. Many were designed to take advantage of picturesque views and stunning natural conditions. The countryside made cheap land available to accommodate larger homes, parks, and other recreational areas. Contrary to the crowded and grim

---

18 Weiss, M.A., *The Rise of the Community Builders. The American Real Estate Industry and Urban Land Planning*, New York, Columbia University Press, 1987.

19 Ruggeri, D., *Death and Life of a Modernist Utopia. The New Town of Zingonia, Italy*, International New Town Institute, Almere, NL, 2010.

20 Merlin, P., *New Towns, Regional Planning and Development*, London, Methuen, 1971.

urban neighborhoods of London, British New Towns would set aside 4 hectares for every 1,000 inhabitants for parks, which were to provide the framework for neighborhood development[21] and an agricultural green-belt that would prevent them from sprawling.

While New Towns are usually associated with high-rise Modernist architecture, in reality many of them featured low-density residential environments of single family or terraced houses. In the UK only one out of five housing units was in high-rise towers, while In France only 30% were single-family homes.[22] This resulted in monotonous physical environments, which resonated negatively in the perceptions of their residents and led to a unique phenomenon called "the new town blues".[23]

From an urban design standpoint, due to the unprecedented size reached by 1960s New Towns, planners sought to embed visual identity or *imageability* into their plans at a variety of scales including the scale of the freeway.[24] Beginning in the mid-1960s walkability – once a key elements in Howard's model – was replaced by the need to accommodate automobile access. This created a mutual dependence that continued to affect New Towns in the decades to come. Milton Keynes in the UK, Almere in the Netherlands and Irvine in the US were all planned with automobile access as a primary concern.

Assuming that most people would choose the car as the preferred means of transportation, planners of these new towns chose to locate of services away from the residential areas, on land clearly visible from the car. To further illustrate how the above conditions of acceptance and rejection have affected the evolution of the New Town ideal, the following pages will focus on the two case studies of Harlow, UK and Irvine, CA, typological examples of the Modernist New Town.

1. Harlow UK as the "Garden New Town"

Planned by Sir Frederick Gibberd on 6,320 acres in pastoral Essex, Harlow was the first British New Town to re-interpret in Modernist forms the healthy balance between town and country envisioned by Howard.[25] The garden city principles adopted by Gibberd included rail

---

21 *Ibid.*

22 Tuppen, J.N., "The Development of French New Towns. An Assessment of Progress", in *Urban Studies*, 20:1, 1983, pp. 11-30.

23 Thomas, W., "New Town Blues", *American Society of Planning Officials, Planning 1964*, Chicago, American Society of Planning Officials, 1964, pp. 184-189.

24 Lynch, K., *Image of the City*, Cambridge, MIT Press, 1960; Appleyard, D., K. Lynch, and J.R. Myer, *The View from the Road*, Cambridge: MIT press, 1964; Ruggeri, D., *Place-Identity, Attachment and Community Attitudes in Suburbia. The Irvine Ranch Case*, Dissertation, Berkeley, CA, University of California, 2009.

25 Gibberd, F., *Harlow New Town*, Harlow, Harlow Development Corporation, 1952.

access, housing variety, the "town park", and a civic center offering residents an urban experience. His adoption of Unwin's "picturesque" approach to urban design was embodied by his use of "street picture". This concept was based on the idea that good urban form would result from the choreographing and manipulation of visual elements. These included "the relationship of house to paving; of form or character, of facade patterns and building lines; the organization of spaces at corners, along roads or at right angles to them, and against more open landscapes".[26]

Figure 4. A Map of Harlow indicating the pedestrian and bike paths serving the New Town

www.harlow.gov.uk

Harlow's plan featured a system of self-contained residential neighborhoods of detached, semi-detached and high-rise residences organized around a small neighborhood park. Each neighborhood featured a shopping center with small local stored and offices located only five minutes from one's home. A system encompassing 100 km of pedestrian and bike trails (figure 4) would offer easy access between the neighborhoods and the industrial areas, with the automobile or transit to link the town center to each neighborhood.[27] Easy access to work would ensure

[26] Jarvis, R., *Urban Environments as Visual Art or as Social Settings? A Review*, in Michael Carmona and Steve Tiesdell (eds.), *Urban Design Reader*, Oxford, Architectural Press, 1980.

[27] Harlow Council, Urban Design Assessment of Harlow, Chapter 4, Final Report Harlow Area Study, April 2005, http://www.harlow.gov.uk/pdf/05%20-%20chapter %204.pdf.

live and work, safer and family-friendly residential neighborhoods. From a landscape standpoint, Harlow featured green wedges, open spaces nestled between two contiguous neighborhoods, often incorporating ecologically sensitive or topographically varied land. In the best British tradition, residents would be given the option to grow their own food in allotment gardens at the edge of each neighborhood. The green wedges would be connected to a large greenbelt dedicated to agricultural, conservation and habitat. In Harlow, the greenbelt incorporated the Stort River, flowing from west to east along the northern edge of the city.

Like many other British New Towns, today's Harlow struggles to address structural issues, such as an aging infrastructure, lack of or limited maintenance of green spaces, unemployment and other social issues, such as some of the highest poverty levels in UK new towns. Nonetheless, its rich landscape and extensive greenbelt continue to serve as resources in terms of habitat and biodiversity.[28] Harlow's unique relationship between the landscape and housing led to a scaled-up and more complex prototype. Such a "Garden New Town", where the landscape is framework for urban development, will become institutionalized in the American New Town of the 1960s and 1970s, which are described in detailed in the following section.

2. *Landscape-driven New Towns in the US*

After World War II, the United States experienced unprecedented pressure to build new homes for a fast growing middle class. Growth occurred at the periphery of American cities, where land was cheap, readily available and easily developable, in socially homogeneous neighborhoods where diversity was prevented by restrictive covenants.[29] New suburban residents were fleeing cities, where crime and poverty were on the rise[30] and settled at the urban edge, in what Robert Fishman has called "bourgeois utopia".[31] Although its impact was limited, New

28 LUC, *Harlow Open Space and Green Infrastructure Study*, Report Prepared for the Harlow Council, 2012.

29 Weiss, M.A., *The Rise of the Community Builders. The American Real Estate Industry and Urban Land Planning*, New York, Columbia University Press, 1987; Brody, J.S., *Constructing Professional Knowledge. The Neighborhood Unit Concept in the Community Builders' Handbook*, Doctoral dissertation, 2009, University of Illinois.

30 Low, S., *Behind the Gates. Life, Security and the Pursuit of Happiness in Fortress America*, New York, 2003, Routledge.

31 Fishman, R., *Bourgeois Utopias. The Rise and Fall of Suburbia*, New York, Basic Books, 1989.

Town development in the US came as a result of legislation promoting master planning and the coordination of the planning process.[32]

The law also gave private corporations access to public funds to plan new communities, which required community builders to coordinate planning decisions with political and administrative partners at the county or state levels.[33]

From a design standpoint, American New Towns deviated from the garden city model in two fundamental ways: their choice of transportation and the use of landscape design as identity shaper. Irvine, Columbia, Reston, The Woodlands and Mission Viejo – the most popular post-World War II American New Towns – did not choose to provide mass transit as their European counterparts. In its place, US developers chose a polycentric auto-oriented urban form. As a result, shopping centers at freeways and arterial intersections replaced the old downtown, often incorporating some of its civic functions. Another element of distinction was the influential role played by landscape architecture with trees and greenbelts framing the streets and announcing each neighborhood to the outside.[34] This was undoubtedly the result of the unique influence of landscape architects like Frederick Law Olmsted Senior and Junior, and John Nolen, whose work was influential in setting the stage for the creation of the American Society of Planners in 1909.[35]

3. Irvine

With a population of 250,000 residents, Irvine is the largest private master planned community.[36] Since its inception in 1961, innovation and experimentation were driving forces in the design of the New Town of Irvine.[37] Innovation was achieved through a collaborative planning process whereby city leaders, developers and design consultants would form "project teams" following a project from its early conception to completion. The City's 1972 "Urban Design Implementation Plan"

---

32 Forsyth, A., *Reforming Suburbia. The Planned Communities of Irvine, Columbia and the Woodlands*, Berkeley, University of California Press, 2005.

33 Weiss, M.A., *The Rise of the Community Builders. The American Real Estate Industry and Urban Land Planning*, New York, Columbia University Press, 1987.

34 Ruggeri, D., *Death and Life of a Modernist Utopia. The New Town of Zingonia, Italy*, International New Town Institute, Almere, NL, 2010; Forsyth, A. and K. Crewe, "A Typology of Comprehensive Designed Communities Since the Second World War", in *Landscape Journal*, 28:1, 2009, pp. 56-78.

35 Peterson, J.A., "The Birth of Organized City Planning in the United States, 1909-1910", in *Journal of the American Planning Association*, 75:2, 2009, pp. 123-133.

36 Campbell, R., "Irvine Back Among Fastest Growing Big Cities Top 10", in *Orange County Register*, 21 June, 2006.

37 Cleland, R., *The Irvine Ranch*, San Marino, CA, The Huntington Library, 2003.

commissioned to Ian McHarg. The plan summarized key urban design principles, flexible enough to accommodate incremental changes in neighborhood planning (figure 5).

Figure 5. Irvine according to Pereira's master plan

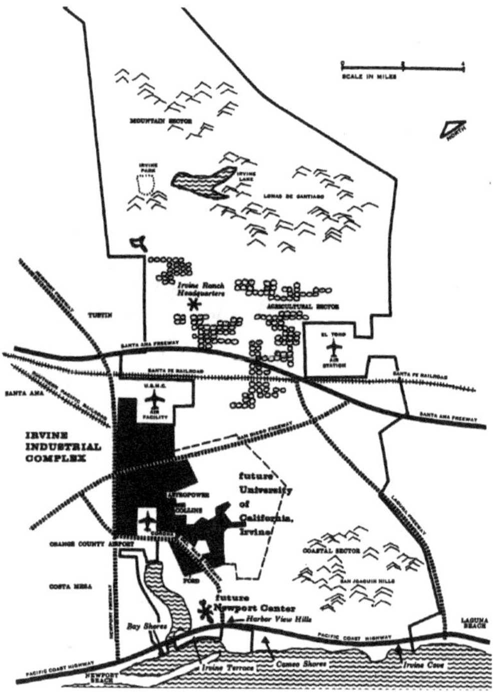

Cleland

Irvine's planning was grounded in the use of Urban Design – first and foremost Kevin Lynch's theory of imageability – to promote a unique identity in contrast to the surrounding sprawl. In Irvine, planners and urban designers concentrated their efforts to the creation of environmental "products". Irvine's houses, shopping centers, and public landscapes were conceived, tested through focus groups and surveys and eventually modified to adjust to changing taste and economics and ensure the perfect fit between the place and its residents.[38]

38 Watson, R., *Planning and Developing the New Town of Irvine, California, 1960-2003: Irvine Company President, 1973-1977; Walt Disney Company Chairman, 1983-1984, an oral history conducted in 2003 by Ann Lage*, Regional Oral History Office, Bancroft Library at the University of California, Berkeley, 2005.

Experimentation continues to drive Irvine's planning, allowing the introduction of New Urbanism principles in the newest neighborhoods, the use native landscaping, and sustainable planning strategies like the recycling of gray water. Whereas other New Towns have struggled to adapt to changing socio economic and environmental conditions, the experimental, landscape-driven process in place in Irvine has become a source of adaptability and resilience. Like Harlow, Irvine has become, in Said's words, the ideal "representation" of the American New Town, dominated by a landscape that shaped its imageability, way-finding, and overall identity. Its institutionalization is witnessed by the numerous attempts to replicate its planning and architectural solutions in countries like Japan, Korea, and China.[39]

D. New Position in Time and Place

Ian McHarg's participation in the master planning of Irvine and later in his master plan for The Woodlands marked the beginning of a new planning trend aimed at the harmonization of natural and human processes in New Towns. Combined with the emphasis on green architecture by architects like Reyner Banham, who saw in green-building technologies a tool to resolve the tension between human needs and ecological health,[40] the work of McHarg prepared the ground for the emergence of the eco-city paradigm (figure 6). Richard Register, founder of the eco-city movement, was among the first to call for the integration of nature in urban areas,[41] together with a new process by which new green communities would be created or retrofitted. However, the true paradigm shift towards "eco-cities" is a very recent phenomenon, which goes beyond the need for green technology to include issues of climate change, governance, and equal access to resources. Sustainable development, bioregionalism, Green design, energy efficiency, and Eco-district development are only of the key principles we are now associating to sustainable cities.[42]

---

39 Watson, R., *op. cit.*; Anton, M., H. Chu and T.S. Writers, "Welcome to Orange County, China", in *Los Angeles Times*, 9, A1, 2002.

40 Banham, R., *Theory and Design in the First Machine Age*, Boston, MIT Press, 1980; Lehmann, S., *The Principles of Green Urbanism. Transforming the City for Sustainability*, Abingdon, Earthscan, 2010.

41 Register, R., *Ecocity Berkeley*, Berkeley, North Atlantic Books, 1993.

42 Roseland, M., "Dimensions of the Eco-City", in *Cities*, 14:4, 1997, pp. 197-202.

Figure 6. Richard Register's vision for an Eco-city

Courtesy ecocity builders

Another important source of change in New Town planning has been New Urbanism, which has promoted the creation of denser, more compact, walkable residential neighborhoods at the urban edge.[43] Most recently, these concerns have expanded to include considerations for health and livability in both cities and suburbs.[44] Also quite new is the emphasis on the quantification of the benefits of sustainable environments, exemplified by the Sustainable Sites initiative benchmarking system. Its goal is to promote sustainable urban design and planning through the preservation of natural resources, energy efficiency, green technology and materials, and sustainable maintenance practices.[45]

Because of their historical link to planning and technological experimentation, New Towns have become fertile grounds for the livable city and eco-city agendas. Efforts to improve energy efficiency and ecological health in New Town are taking place in East Asia, North Africa and the Middle East. Windmills, solar panels, geo-thermal plants, natural and constructed wetlands have become part of the landscape vocabulary

---

43 Calthorpe, P., *The Next American Metropolis. Ecology, Community, and the American Dream*, New York, Princeton Architectural Press, 1995.

44 Southworth, M. and P.M. Owens, "The Evolving Metropolis. Studies of Community, Neighborhood, and Street Form at the Urban Edge", in *Journal of the American Planning Association*, 59:3, 1993, pp. 271-287.

45 Sustainable Sites Initiative, *The Sustainable Sites Initiative. Guidelines and Performance Benchmarks 2009*, http://www.sustainablesites.org/report.

of contemporary New Towns, and represent the most recent evolution of a trajectory that begun with Ebenezer Howard's garden city.

This stage in the evolution of the New Town ideal presents great opportunities and challenges to planners and designers worldwide. While technology may in fact reduce the environmental impact of New Towns, it should not be seen as "the" ultimate goal. In Howard's Garden City, as in the New Towns of the early and late 20th century, technology was a means to an end, which was the improvement of the quality of life for an increasing number of urban dwellers through increased access to services, better jobs, and healthier environments. Today, the emphasis on technological solutions to sustainability has taken center stage in the work of environmental designers, overshadowing concerns for equity, democracy, and responsiveness of urban landscapes to the needs for beautiful, livable places. They have become institutionalized in the policies of many governments worldwide. Said's "new position in time and place" in the evolution of the New Town ideal is illustrated by the recent efforts to create the ecological New Towns of the 21st century, like the Sino-Singapore Eco-city and the New Town of Sidi Abdellah, in Algeria, which are discussed in detail in the following pages.

1. Sino-Singapore Tianjin Eco-City

The Sino-Singapore Eco-city in Tianjin is the poster child of eco-city development born out of a strategic cooperation between the governments of China and Singapore. Its goal is to serve as a model for the livable city of the future, incorporating the type of comprehensive, coordinated and green solutions required to strike a healthy new balance between nature and urbanization. Over the next 30 years, half a billion Chinese residents are expected to move from the countryside to cities.[46] In response to such projections, the country has begun to plan dozens of New Towns for which sustainability has become a key planning principle. Through this process, the traditional definition of New Town has evolved to incorporate "triple bottom line" green urbanism based on zero fossil-fuel energy use, zero waste, and low-to-no-carbon emissions,[47] and the eco-city projects currently under way are ideal representations of this transformation.

---

[46] Harvard International Review, http://hir.harvard.edu.

[47] Lehmann, S., *op. cit.*

Figure 7. Master plan for the 25 square km Tianjin Eco-city

Courtesy: Sino-Singapore Tianjin Eco-city Investment & Development Co. Ltd.

Located a few hours south of Beijing and 45 km from Tianjin's city center, the Sino-Singapore Eco-city is the first of a new generation of New Towns, focused on reducing China's ecological footprint by conceiving of an energy efficient, sustainable urban form for a future projected population of 400,000 new residents. The Eco-city encompasses an area of 31 square kilometers of dry, scarcely vegetated, polluted land in what would otherwise be a very sensitive eco-tone between land and sea (figure 7). It aims to create an ideal new urban environment combining livable neighborhoods, green industry, and natural areas reclaimed using innovative remediation technologies. Other important features include a mandate for green buildings, clean and renewable energy, and efficient public transit within 500 meters of every new home.

In an effort to move beyond the notion of sustainability as technological fix, Sino-Singapore Eco-city planners have tried to address issues of equity and access by requiring that 30% of the total housing be affordable, and that basic social services be provided regardless of one's income. The displacement of existing communities of fishermen and farmers, assumed to be an inevitable price to pay for change, has been offset by the promise of employment in one of the new "green" factories. Local residents have not been engaged in the design and planning of the Tianjin Eco-city, a departure from some of the European precedents that inspired it. Furthermore, little consideration has been given to educating the public about the benefits, challenges and responsibilities of sustainable living.[48] Marketing brochures showcase glass skyscrapers green roofs, and urban parks devoid of human activity. Unfortunately, the shortcomings of this project are quickly being transferred to the national level, where the Eco-city has become a model for many similar communities.

2. *Sidi Abdellah*

Sustainable development is the theme underlying the planning of Algeria's most recent New Town of Sidi Abdellah. The city occupies 7,000 hectares of pastures and agricultural fields located 45 km from the center of Algiers (figure 8). Like the Swedish and French New Towns of the 1960s and 1970s, from which it draws inspiration, Sidi Abdellah is not planned as an alternative to Algiers, but it is dependent on the capital city for most major services. This is reflected in its location, chosen for its proximity to a national highway and the potential for a future railroad connection to the nearby capital.

Through the implementation of Algeria's first "eco-smart" New Town, the Algerian Government wants to be recognized as a key player in advancing the world's agenda on sustainable development. Its interest in sustainability is reflected in the project goals indicated by the European New Town Platform.[49]

1. To reduce coastal development and sprawl
2. To re-direct growth and development to designated sites;
3. To reinforce and improve the attractiveness of the nation's territory;
4. To envision new urban growth based on knowledge economy and information
5. To promote urban design and architectural excellence.

Similar to many other world New Towns, Sidi Abdellah's plan places great emphasis on a system of open spaces that includes a large central park and a "protection" greenbelt surrounding the New Town.[50] The

---

48 *Ibid.*

49 European New Towns Platform, Préparation du 2e panel d'experts de Novembre 2011, Projet de Ville Nouvelle de Sidi Abdellah, unpublished manuscript, 2011.

50 *Ibid.*, p. 17.

landscape plays a double role as both a unifying element and a buffer between each residential neighborhood and the surrounding industrial and commercial areas and it is an integral part of the vision for the New Town. Nonetheless, a number of questions may be raised with regard to the 3,000 hectares open space provision. First and foremost, how will this land be maintained and used? Secondly, it is unclear how the local administrations will deal with pressure to encroach onto the greenbelt, given that no regulations have been put in place to that end.

Figure 8. Master Plan for the New Town of Sidi Abdellah

European New Town Platform

An important aspect of Sidi Abdellah's is the role played by technological innovation and education in establishing a unique identity for the New Town. This is important, given that one of the goals has been to create a "competitive cluster and attract long-term industrial investment".[51] In response to the need for innovation, an infrastructural "spine" has been envisioned that runs in a tunnel underneath major arterial roads. Such spine includes all major infrastructure necessary to both residential and business uses, including electricity, water, sewage, gas, fiber optics, and water.

According to some of the experts consulted during a "panel of experts" organized there in November 2011, Sidi Abdellah's plan does not seem to integrate social sustainability principles. The same International experts have recommended that residents of the existing hamlets of

[51] *Ibid.*, p. 8.

Zéralda, Douéra, and Souidania be included in the planning process.[52] The approved master plan isolates pre-existing settlements and newly planned neighborhoods by using open space and high-speed arterial roads as buffers between them. The expert's feedback included the suggestion to look into the redesign of those areas to better integrate old and new settlements and create a more cohesive community.[53]

III. Observations on a Traveling Ideal

How has the New Town ideal weathered 200 years of history? What elements have been consistently present in the planning of garden cities, New Towns, and Eco-cities? Is it possible to speak of trends in the evolution of such ideal? And what may be the risks and opportunities connected with the adoption of the New Town ideal that has traveled to us?

This research effort, which included a survey of 10 of the most popular New Towns, reveals a dual pattern in the choice of locating New Towns, related to the goals and objectives underlying their developments. The diagram in figure 9 shows a) that most New Towns are located an average of 40 miles (64 kilometers) from the metropolitan areas they serve, true to Howard's model of a garden city in close proximity to the historic city; b) That despite existing plans to connect eco-cities through high-speed trains in China and North Africa, the future accessibility of these communities is questionable. Today's Eco-cities have not yet built such transportation systems, which may reflect negatively on their ability to offer truly sustainable lifestyles, especially in the presence of an uncertain economic future, which may not allow funding such expensive infrastructure.

52 European New Towns Platform, Project New Medina. Second Experts' Panel Report, unpublished manuscript, Sidi Abdellah, Algeria, November 21th-23rd, 2011.

53 *Ibid.*

Figure 9. Diagram showing relative distance from their respective referent cities and type of transportation linkages between them

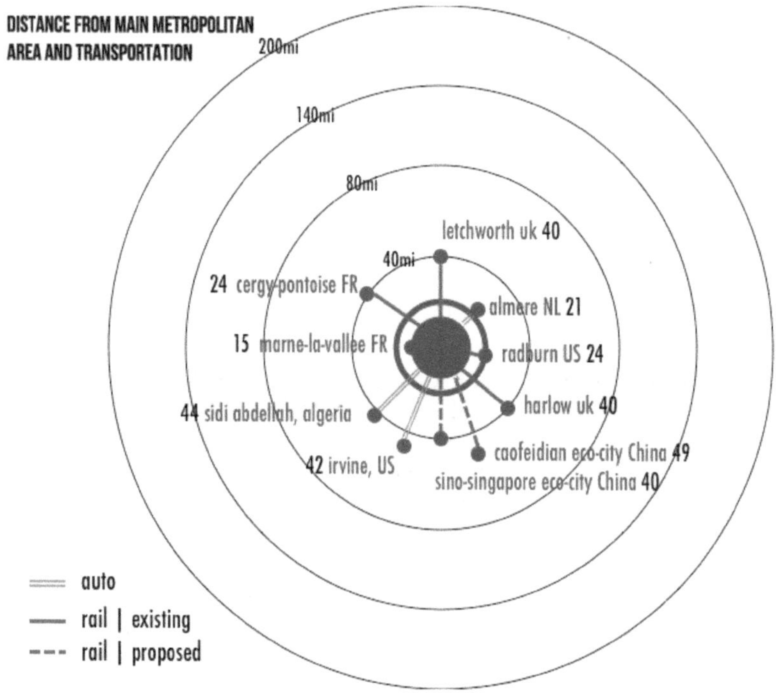

Illustration by author

Our comparison also raises the issue of the ideal size of New Towns, which have grown from the 32,000 people envisioned by Ebenezer Howard to the 400,000 inhabitants expected to move into Tianjin's eco-city. The chart in figure 10 illustrates the extent of such growth, and points at the 1960s as a watershed moment in the evolution of the New Town model. The reason can be found in the diffusion of the automobile, which gave those communities the freedom to expand both physically and socially, while weakening their healthy reliance on efficient public transit. Without a provision for public transportation, a key feature in Howard's model is getting lost.

Figure 10. Total Population of a sample of New Towns. In light grey, the projected growth

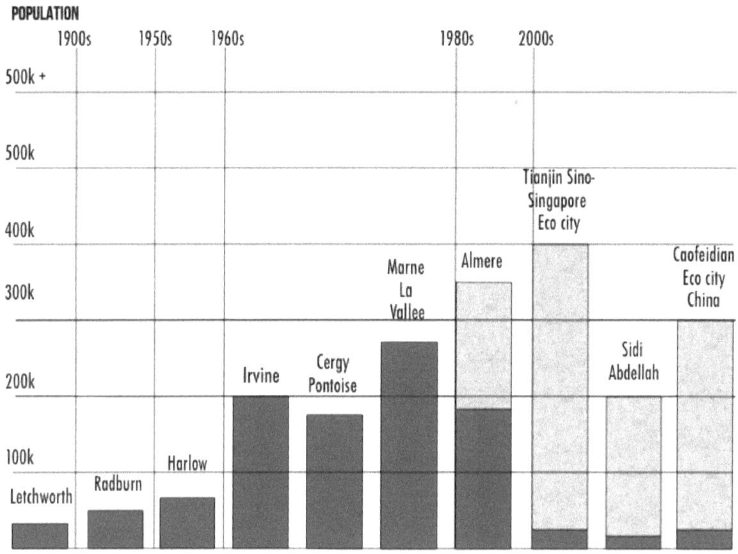

Illustration by author

The increase in size of New Towns opens a discussion on the neighborhood as the intermediate scale between home and city. In the US after World War II, Perry's Neighborhood Unit became the institutional model used to plan and size a residential neighborhood. It was based on the notion that schools would act as social hubs and that 5,000 people were needed to support a local elementary school. The "eco-district" concept has become synonymous with neighborhood in the context of the current sustainable development trends. It interprets energy efficiency and self-reliance in sociological terms, by sizing neighborhoods based on the energy produced and consumed locally. This interpretation may become problematic, as it does not correspond to the residents' perceptions and experience of their own community.

Figure 11. Population gross density expressed in people per square kilometres

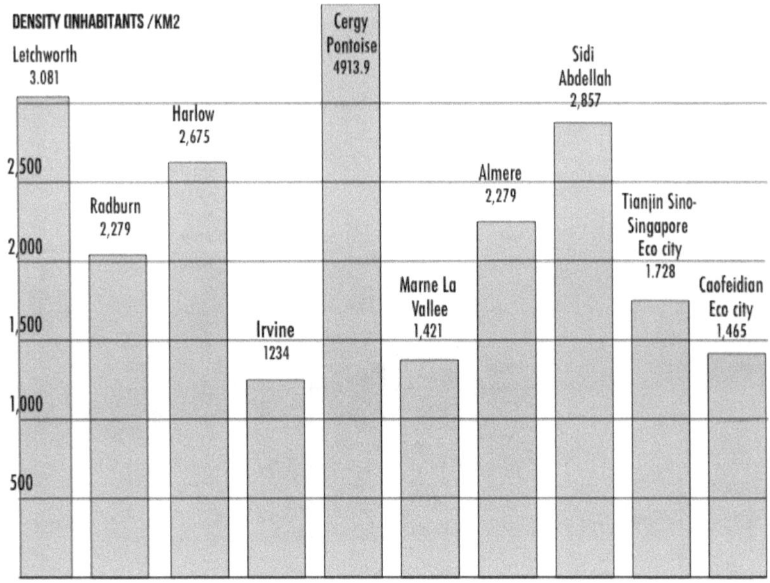

Illustration by author

Another aspect that emerges from our analysis is an inconsistent approach to density, with a generalized decline in the density per square kilometer from the original garden cities in Europe and the US (figure 11). The Chinese eco-city of Tianjin shows densities similar to suburban, auto-oriented New Towns such as Irvine, Almere and Marne-la-Vallée. This could potentially be fatal to the long-term resilience of these communities. Low densities of 7 to 8 dwelling units per acre may in fact be insufficient to support the vibrant neighborhood environments of the sustainable city.[54]

The variability in population density illustrates the range in acceptance and rejection of the New Town model and its character as a Traveling Theory, which Edward Said describes as an idea shaped by the contextual conditions at the time of its adoption. Modernist new towns, particularly those created by private efforts have, with their low densities, contributed to the institutionalization of a New Town image as a low density environment, a character which has since then consistently been present from World War II to nowadays' eco-city.

[54] Lehmann, S., *op. cit.*

Figure 12. Number of acres devoted to open space, shown as dark grey if inclusive of a greenbelt

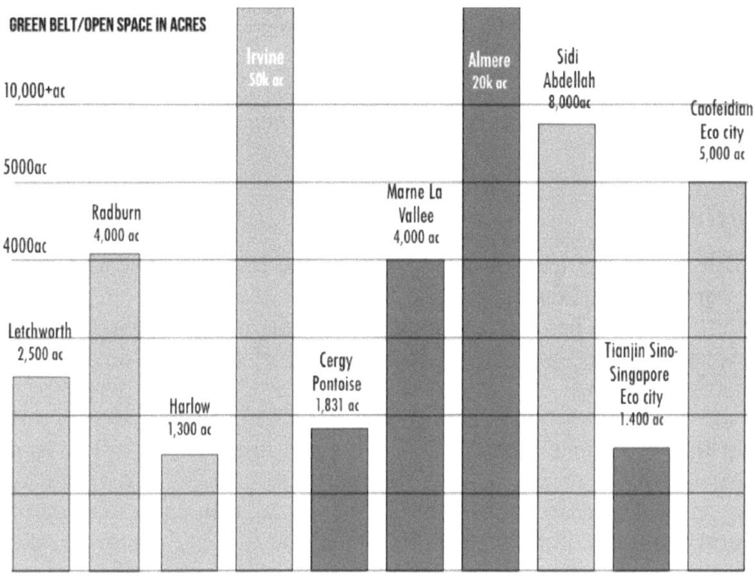

Illustration by author

One last set of considerations has to do with the open space allocation in the planning of New Towns. Figure 12 shows the amount of open space available to residents in the form of parks, greenbelts, and ecological corridors. Overall, open space figures have increased dramatically overtime. This may be ascribed to the increase in the total population of New Towns, with the subsequent request of additional open space to support public health, leisure, and physical activity. This is particularly important given the growing concerns for obesity, cardiac diseases, and other health issues related to the lack of physical activity. Nowadays 65% of the world's population lives in countries where obesity causes more deaths than malnutrition and 40 million children suffer from obesity.[55] Researchers have found positive links between availability of open space within walking distance of one's residence and human health and stress reduction but open space availability and size must be carefully weighed against the need for higher densities as a necessary requisite for sustainable living.

[55] WHO, "Obesity: Preventing and Managing the Global Epidemic", World Health Organization technical report series, 2000, p. 894.

Another reason for the increase of open space areas in New Towns may be the influence of the environmental movement on their planning, and the evolution of the garden city to include an increasingly larger amount of habitat areas and storm water management and other forms of eco-system services.[56] This represents a contemporary refinement of the more general balance between country and city included in Howard's Garden city model, with the landscape assuming a more central and integral role in the lives of New Town residents, their communities, and nations.

Conclusions

This paper has tried to shed light on the historical and theoretical roots of the New Town ideal and to explain how the travelling of this theoretical model has led it to change and adapt to evolving socio economic, cultural and geographical conditions. What has become obvious is that while no researcher has yet found an acceptable common definition of New Town, such definition should be operational and open, rather than fixed and unchangeable. First and foremost, New Towns are *comprehensive* attempts at providing not only housing, but also a new lifestyle for the urban masses displaced by wars, economic restructuring or rural to urban migration affecting many developing countries nowadays. While past New Towns show a moderate degree of success in shaping New Town with strong identities and resilient economics, a key question for the future is whether these models will be able to succeed given the unprecedented scale and motivations of today's urban immigrants.

Another important element of disjuncture in the planning of New Towns has been the relationship with their metropolitan areas, which resulted in two alternative models. On one hand is the synergistic approach exemplified by Swedish, Dutch and French New Towns, whereby new communities relied on the historic city for access to services and cultural activities. On the other hand is a "competitive" approach that allowed a healthy independence of New Towns from the old city and the emergence of a "new" urban identity as in the British and American models. The latter is more conducive to the eco-city idea, since it supports the principles of compactness, walkability and proximity to urban services. This highlights the priority of the issues of livability, density, mixed use, and sustainable transportation over idealized relationships between new and old city.

---

56 Lehmann, S., *op. cit.*

An important issue emerging from the contemporary discussion on eco-cities is that of the ideal size of a neighborhood. Size matters to the functioning of a neighborhood. While Perry's neighborhood unit may need to be adapted to the changes in socio-economic makeup and life-style prompted by the information society, we should be critical of energy-based definitions of community based on energy production, rather than social capital construction. This also underscores the importance of the garden/landscape as both a physical and social framework for community, regardless of which country, socio-economic or cultural context we are working within. And it is from this landscape that we should begin to envision the "eco" New Town of the future.

Bibliography

Almere, http://english.almere.nl/local_government/almereprinciples.

Appleyard, D., K. Lynch and J.R. Myer, *The View From The Road*, Cambridge, MIT press, 1964.

Anton, M., H. Chu, and T.S. Writers, "Welcome to Orange County, China", in *Los Angeles Times*, 9, A1, 2002.

Banham, R., *Theory and Design in the First Machine Age*, Boston, MIT Press, 1980.

Birch, E.L., *Radburn and the American Planning Movement. The Persistence of an Idea*, New York, Taylor & Francis, 1980.

Brody, J.S., *Constructing Professional Knowledge. The Neighborhood Unit Concept in the Community Builders' Handbook*, Doctoral dissertation, University of Illinois, 2009.

Calthorpe, P., *The Next American Metropolis. Ecology, Community, and the American Dream*, New York, Princeton Architectural Press, 1995.

Campbell, R., "Irvine Back Among Fastest Growing Big Cities Top 10", in *Orange County Register*, 21 June, 2006.

Cleland, R., *The Irvine Ranch*, San Marino, CA, The Huntington Library, 2003.

Corbusier, L., *The City of Tomorrow and Its Planning*, London, Dover Publications, 1987.

Corner, J., "A Discourse on Theory I: 'Sounding the Depths' – Origins, Theory, and Representation", in *Landscape Journal*, 9:2, 1990, pp. 61-78.

Cosgrove, D., *Social Formation and Symbolic Landscape*, Madison, Wisconsin, University of Wisconsin Press, 1983.

Doglio, C., *La Città Giardino*, Roma, Gangemi, 1985.

Duvall, R. and L. Varadarajan, "Traveling in Paradox. Edward Said and Critical International Relations", in *Millennium – Journal of International Studies*, 36:1, 2007, pp. 83-99.

ENTP, European New Towns Platform, http://www.pilotcities.eu.

European New Towns Platform, *Préparation du 2e panel d'experts de Novembre 2011*, unpublished manuscript, 2011.

European New Towns Platform, *Project New Medina. Second Experts' Panel Report,* unpublished manuscript, Sidi Abdellah, Algeria, November 21th-23rd, 2011.

Fishman, R., *Bourgeois Utopia. The Rise and Fall of Suburbia*, New York, Basic Books, 1989.

Forsyth, A. and K. Crewe, "A Typology of Comprehensive Designed Communities Since the Second World War", in *Landscape Journal*, 28:1, 2009, pp. 56-78.

Forsyth, A., *Reforming Suburbia. The Planned Communities of Irvine, Columbia and The Woodlands*, Berkeley, University of California Press, 2005.

Gaborit, P., *European New Towns. Image, Identities, Future Perspectives*, Brussels, Peter Lang, 2010.

Gibberd, F., *Harlow New Town*, Harlow, Harlow Development Corporation, 1952.

Guardian, "Top Tips: Garden Cities and New Towns", http://m.guardian.co.uk//housing-network/2012/dec/08/top-tips-garden-cities-and-new-towns.

Harlow Council, Urban Design Assessment of Harlow, Chapter 4, Final Report Harlow Area Study, April 2005, http://www.harlow.gov.uk/pdf/05%20-%20chapter%204.pdf.

Harvard International Review, http://hir.harvard.edu.

Howard, E., *Garden Cities of To-Morrow. A Peaceful Path to Real Reform.* London, Swan Sonnenschein & Co. Ltd., 1902.

International New Town Institute, INTI, Almere, the Netherlands, http://www.inti.net.

Jarvis, R., "Urban Environments as Visual Art or as Social Settings? A Review", in Michael Carmona and Steve Tiesdell (eds.), *Urban Design Reader*, Oxford, UK, Architectural Press, 1980.

Lehmann, S., *The Principles of Green Urbanism. Transforming the City for Sustainability*, Abingdon, Earthscan, 2010.

Low, S., *Behind the Gates. Life, Security and the Pursuit of Happiness in Fortress America*, New York, Routledge, 2003.

LUC, "Harlow Open Space and Green Infrastructure Study", Report Prepared for the Harlow Council, 2012.

Lynch, K., *Image of the City*, Cambridge, MIT Press, 1960.

Merlin, P., *New Towns, Regional Planning and Development*, London, Methuen, 1971.

NCDC Urban Practitioners, *Northern Bedfordshire Urban Design assessment. Letchworth Garden City*, London, NCDC, 2007.

Osborn, F.J. and A. Whittick, *New Towns. Their Origins, Achievements and Progress*, London, L. Hill, 1977.

Peterson, J.A., "The Birth of Organized City Planning in the United States, 1909-1910", in *Journal of the American Planning Association*, 75:2, 2009, pp. 123-133.

Projet de ville nouvelle de Sidi Abdellah. Prepared for the European New Medina Project, retrievable at http://newmedina.eu.

Register, R., *Ecocity Berkeley*, Berkeley, North Atlantic Books, 1993.

Roseland, M., "Dimensions of the Eco-City", in *Cities*, 14:4, 1997, pp. 197-202.

Ruggeri, D., Place-Identity, Attachment and Community Attitudes in Suburbia. The Irvine Ranch Case, Dissertation, Berkeley, CA, University of California, 2009.

Ruggeri, D., *Death and Life of a Modernist Utopia. The New Town of Zingonia, Italy*, International New Town Institute, Almere, NL, 2010.

Said, E.W., *'Traveling Theory'. The World, the Text, and the Critic*, Cambridge, Harvard University, 1983.

Short, John R., "Urban Policy and British Cities", in *Journal of the American Planning Association*, 48:1, 1982, pp. 39-52.

Sidi Abdellah, http://www.anpt.dz.

Sitte, C., *City Planning According to Artistic Principles*, New York, Rizzoli, 1986.

Southworth, M., and P.M. Owens, "The Evolving Metropolis. Studies of Community, Neighborhood, and Street Form at the Urban Edge", in *Journal of the American Planning Association*, 59:3, 1993, pp. 271-287.

Stein, C.S., "Toward New Towns for America", in *The Town Planning Review*, 20:3, 1949, pp. 203-282.

Stewart, D.J., "Cities in the Desert. The Egyptian New-Town Program", in *Annals of the Association of American Geographers*, 86:3, 1996, pp. 459-480.

Sustainable Sites Initiative, *The Sustainable Sites Initiative: Guidelines and Performance Benchmarks 2009*, 2009, available at http://www.sustainablesites.org/report.

Thomas, W., "New Town Blues", in *American Society of Planning Officials, Planning 1964*, Chicago, American Society of Planning Officials, 1964, pp. 184-189.

Tianjin Sino-Singapore Eco-City, http://www.tianjinecocity.gov.sg.

Tuppen, J.N., "The Development of French New Towns. An Assessment of Progress", in *Urban Studies*, 20:1, 1983, pp. 11-30.

Watson, R., *Planning and Developing the New Town of Irvine, California, 1960-2003: Irvine Company President, 1973-1977; Walt Disney Company Chairman, 1983-1984, an oral history conducted in 2003 by Ann Lage*, Regional Oral History Office, Bancroft Library at the University of California, Berkeley, 2005.

Weiss, M.A., *The Rise of the Community Builders: The American Real Estate Industry and Urban Land Planning*, New York, Columbia University Press, 1987.

WHO, "Obesity: Preventing and Managing the Global Epidemic", *World Health Organization technical report series*, 2000.

Informal housing in Cairo © Edoardo Guglielmetti

Informal housing in Cairo © Pascaline Gaborit

HBRC site construction, eco building © Edoardo Guglielmetti

New Fayoum © Pascaline Gaborit

New Fayoum, house from the project « Ibni Beitak – build your own house » © Edoardo Guglielmetti

Traditional Buildings (Nubian architecture) adapted to climate, Tunis Village, Fayoum region © Edoardo Guglielmetti

Fayoum Surroundings © Edoardo Guglielmetti

ALGERIA

Sidi Abdellah © Edoardo Guglielmetti

Sidi Abdellah © Pascaline Gaborit

Sidi Abdellah © Pascaline Gaborit

Jean-Jacques Luz buildings in Mahelma village, site of Sidi Abdellah © Edoardo Guglielmetti

MOROCCO

Wind turbines and existing village around Chrafate © Mohamed Khandriche

Chrafate © Mohamed Khandriche

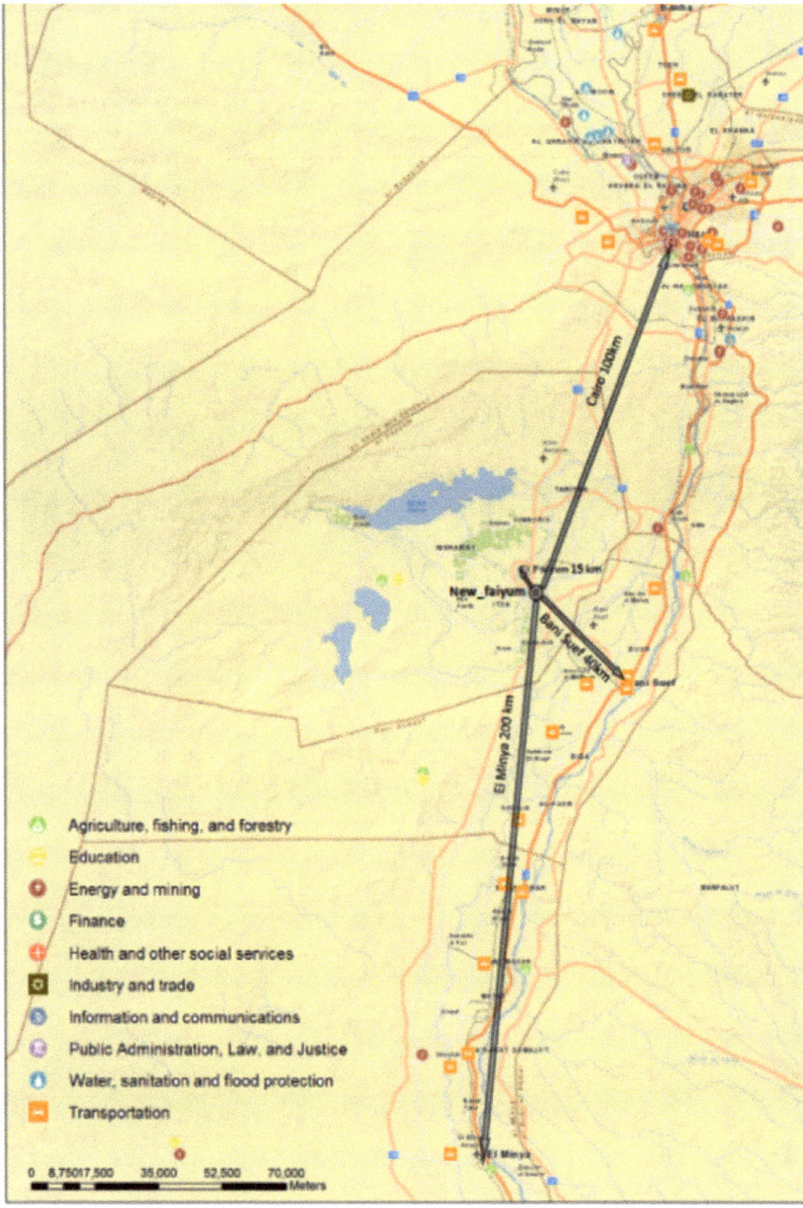

New Fayoum Location © HBRC

Master plan New Fayoum © NUCA

Building Sustainable Desert Towns for Citizens

Christian HORN

Architect and Urban Planner

Recomposing the Stakeholders' Relationship

The social and political changes in Egypt since the revolution in 2011 have strengthened the position of a new stakeholder in housing construction, which until now has been neglected by public authorities. Let's call him "the citizen". The recent events have given attention and a voice to the people, struggling for a better life under better conditions. While in Egypt public authorities have primarily dealt with project developers, construction companies, institutions and influential individuals in planning and constructing new towns, they now have the chance and the necessity to listen to the population, including the ordinary citizen, and to take into accounts their needs and demands.

Major improvements that could and should be done for the development of new towns, such as New Fayoum for instance, depend on the interaction between the public authorities and the new stakeholders. The time of arbitrary, sweeping relocations from informal settlements should be over. One of the big steps for the public authorities to accomplish is to meet, understand, and involve the citizens through a shared learning process. A process that is still and probably always will be a big challenge in all countries with democratic systems. Recent examples in Europe are the confrontations between citizens and public authorities in 2010 on the extension of the main train station in Stuttgart, Germany, or the on-going confrontations about the new airport for western France in Notre-Dame-des-Landes. Those examples have first to be analysed in detail to understand the exact reasons of these confrontations and the balance of power, but they show us the difficulties and conflicts in these shared learning process.[1]

1 Similar examples in France are analysed by Philippe Subra in his work "Géopolitique de l'aménagement du territoire".

In this new landscape of relationships between the stakeholders, the public authorities must also readdress their relationship with project developers. If developers may remain irreplaceable in the construction process, their actions must be reviewed, controlled, and redirected if necessary. Developers' interests and those of their shareholders are not always similar to those of the public authorities and the citizens. These various parties may hold different priorities, different time schedules, and different concepts of return on investments. Since construction projects are particularly vulnerable to corruption and illegal arrangements, special attention must be given to an independent control of commissions and lobbying. The NGO Transparency International has developed a series of tools to prevent corruption on construction projects.[2]

Until now the process of developing entirely new towns in Egypt seems to be a successive and sequenced intervention of the different stakeholders, with limited interactions and an unbalanced relationship between them. The opinions of project developers, construction companies, institutions and influential individuals were favoured in this process, while the voices of the local population went unheard. Taking into account the new social and political landscape, the development of new towns should be transformed into a process of continuous collaboration and interaction, building a balanced relationship between the main stakeholders during the whole development process.

Erbil Citadel, Iraq Kurdistan

© Gordon 2007

The Fayoum Oasis

© author 2012

Public authorities have to guarantee the respect of the public interest from the beginning of the process while other stakeholders are successively integrated in the project. Establishing and maintaining this balance is a complicated but necessary task for sustainable development and needs independent and well skilled institutions.

2 Transparency International, http://www.transparency.org.

I. Why Build a City in the Desert?

Building in the desert is an important issue for the future of Egypt and other desert countries. Egypt will have an estimated population of 123 million inhabitants by 2050.[3] As the population is today about 82 million residents, it will need to provide housing for 40 million more inhabitants within 40 years, an incredible challenge. The metropolis of Cairo might have reached the 20 million inhabitants today and the population and houses keep on growing on the fertile land of the Nile Delta.

New Gourna Mosque

© Marc Ryckaert

Searching for examples of contemporary desert towns, I only found few examples. The older communities, such as New Gourna Village, planned by the Egyptian architect Hasan Fathy and constructed between 1945 and 1948, are big sources of inspiration on a smaller scale. Today's fast-growing towns in some Arab countries, such as Dubai and Abu Dhabi, are fascinating case studies, but they're built with substantial financial investments that are not available for New Fayoum. For instance, Masdar City in the United Arab Emirates, designed by the British architect Norman Foster and currently under construction, is an inspiring project. However, it is too expensive to be followed closely for

3 United Nations, Department of Economic and Social Affairs, Population Division, http://www.un.org/esa/population/unpop.htm.

Egypt's upcoming communities. Still, Masdar and other examples should be analysed, as some ideas could be adapted to the local context, and some of today's cutting-edge technology might become more accessible and affordable in the near future.

One of Egypt's important challenges is to develop a sustainable model of new desert cities for a population with limited financial means and partly coming from rural areas. Most of the recently built housing is too expensive for a large part of the population, and the state will not be able to subsidize affordable housing in the quantities needed.

Egypt will need research, innovation, planning, and cooperation to develop a general model or framework, which can then be adapted to specific sites, local climates, and future populations. Applying generic master plans, without tailoring them to the geographic, climatic, cultural, and social contexts of each respective location, would not be successful.

We saw the importance of understanding, respecting and taking into account the site context in the new town of Sidi Abdallah in Algeria.[4] Planned on a rather hilly area the integration of the local topography is essential for issues like transport planning and infrastructure, rainwater and erosion, identity and image of the new town. Topography influences many aspects of an urban structure, like the layout and use of the different means of mobility (implementation of tramways, use of bicycles, walking habits), the densities and the building heights (enhancing the higher points of the area) and the building itself (view on the landscape, natural ventilation, shades). It is also an important feature to create an identity and strengthen links between the new communities.

As these will be entirely new towns, communities just beginning their history, some essential questions must be examined from the beginning to integrate future issues in the new towns' structure. Copying today's urban structures, which are heavily reliant on cheap fossil energy, would not be a good starting point. Instead, we might have a look on the approaches used in Masdar City to develop a new view on cities.

Just as an example, in regarding mobility some questions could be: how many inhabitants will be able to afford using cars with combustion engines in 20 years, when petrol prices will be double or triple their current levels? Are solar power and electric cars an alternative for desert towns and will they be affordable for everybody? Or does the end of cheap fossil energy spell the end of the individually owned car? If car-based transportation recedes, will mixed used local neighborhoods play an essential role in the layout of new towns and the life of people? How will street landscapes and their uses change? Answering these questions

4 The expert panel on the new town of Sidi Abdellah took place between the 21/11/2011 and 24/11/2011.

will have a large impact on any new town's structure and its integration in its environment.

To tackle important challenges such as this one, an innovative and flexible research centre, with a small overhead, should start rapidly to develop international partnerships and contextualize at a local level, to nourish the work and discussions on New Fayoum and other desert cities. Egypt is far from alone in this need. On the contrary, Egypt could become a mayor centre of innovation in this task. As the world population continues to grow, both in rural and urban areas, other desert countries are currently facing similar questions – and through climate change, some of today's existing towns in semiarid climate zones might become desert towns in future. One example of similar research is the Young Cities project, based on the case study of Hashtgerd new town in Iran, situated 65 kilometres west of Tehran. It was planned for a target population of 500,000 inhabitants, but has only about 30,000 inhabitants today.[5]

Masdar City

© Jan Seifert

[5] "The project's overall ambition is to develop, implement, and evaluate strategies, technologies, and methodologies for sustainable, energy-efficient, and resilient urban settlements in arid and semi-arid regions, especially the Tehran-Karaj region. (Source: http://www.youngcities.org)".

II. Transparency in the Decision-Making Process

Seeing a city like New Fayoum for the first time raised an essential question valid for the development of new towns in the desert (like New Fayoum in Egypt or Boughezoul in Algeria): Why build a city here? Historically, cities were founded primarily for geographical reasons: a crossing of trade routes, a river or port, a territory with fertile land, or a strategic area for defence, for example. Well-situated cities on the lucky side of history had significant advantages. Some modern cities have had centuries to develop at a much slower pace than Egypt's new towns will experience.

There are surely reasons to build at a specific site, but these motives do not seem to be linked to economy, agriculture, or defence. Instead, the reasons seem to hinge on political, administrative, or financial influences.[6] But those "hidden" reasons are less convincing and often do not assure the good development of the town, in the readily apparent way that geographical advantages do. It seems the local population does not trust the intentions behind these decisions, and they express this by ignoring the partly constructed new towns and by not settling down there. In the future the Egyptian state authorities have to answer the question: Why should people move to a desert town and what will life be like in there in ten years' time?

Egypt for instance, is a highly centralized country. In 1996 Cairo concentrated 43% of the public employees, 40% of the private employees and 40% of the industrial employees. Also half of the universities of Egypt were concentrated in the capital city.[7] So how can you ask people to move to a new town, when the employment opportunities or education facilities do not follow? Creating or relocating important public facilities in the new towns from the beginning is a strong gesture that the state itself is convinced and believes in the success of these towns. This goes along with the development of an efficient public transport system between those new centralities and leads slowly to decentralization. Decentralization, that will beneficial and necessary for Egypt and might lead to a more balanced development of the country.

---

6 Officially New Fayoum was established by presidential decree in 2000 to accommodate a population of 120,000 inhabitants.

7 Barthel, P.-A., "Relire le Grand Caire au miroir de la densité, Confluences Méditerranée", No. 75, autumn 2010; quoted in the book of Jean-Pierre Sueur *Villes du futur, futur des villes*.

Busy street life in Cairo

© author 2012

New Fayoum

© author 2012

The process of decision-making must change in order to address the missing answers to such essential questions. Establishing transparency in decisions, debates, and understandable arguments is necessary to attract future inhabitants and to get valuable feedback from the population about the strategy of the project. Equally, methods of communication must adapt and strong gestures from state authorities might also be necessary to assure and invest in the promising future of these towns.

This should also be seen in the long history and experience of Egypt with new town development that started in 1977 with the Tenth of Ramadan City. Projects in this first generation of new towns did not often succeed and some of them are still in development. Some of the French new towns, created in the 1960s, suffered in the beginning as the French state hesitated to create or relocate public facilities. As an example, the new town of Cergy-Pontoise, constructed from 1970 onwards, had to wait 20 years until 1990 before becoming a university town with today between 13,000 and 14,000 students.[8]

The project of New Fayoum and other third generation new towns can and should play a leading role in a new way of working with an emphasis on constructive discussion and transparent decisions – a necessary step to bring New Fayoum and other new towns to life.

Constructing in the Fayoum Oasis

© author 2012

[8] Warnier, B., *Cergy-Pontoise du projet à la réalité: Atlas commenté*, Sprimont, Mardaga, 2004.

III. Starting with a Community

Living towns are not simply built on concrete foundations but on the human need for interaction. The actual master plans of some new towns with their separated zones (residential, service, tourism, and so on), does not illustrate the complexity of human interaction. People move to towns to find new and better opportunities to work, and to share their lives with others. They hope that in a busy community, where many people are living and working together, they will find a service to offer, a product to sell, and a customer base. They hope to integrate or build up a local community for support, shared experiences, and guidance.

Such opportunities exist in a mixed town, where people work close to their living spaces, where shops are situated in ground floors of office buildings, and where small factories are inside dense urban fabrics, with all the conflicts that arise out of this proximity. It cannot be achieved through large-scale zoning and mono-functional areas. In France today, we see problems resulting from the large housing estates built in the 1950s and until the 1970s, and we anticipate similar issues with the mono-functional residential zones on urban peripheries as a consequence of the economic crises and rising transport costs. So an easy rule to follow should be the creation of at least one local job for every family moving to the new town.

The building process itself seems to be splintered, with tasks that are separately planned and executed. While the construction of technical infrastructure and streets is far advanced, and housing for about 18,000 people has been built in one residential zone, other spaces for offices, shops, and leisure and health facilities are not finished or not in service. But to welcome the first families arriving, the new town needs all basic services so that people can settle in comfortably: places to work, to learn, to buy groceries, to relax, and to observe religious practices. Of course, the first residents need a basic technical infrastructure, but they do not need an entire infrastructure for some 100,000 inhabitants. These families need an open bakery, not roads leading to empty plots.

The construction process should not advance in consecutive layers (roads, green spaces, housing, industry, and so on), but instead should start with a well-planned core, where the main elements of a new town are built and begin service. A central square with transport facilities, a town hall, a market, a restaurant, a place of worship – these ingredients establish a local community that people can identify with and trust. Residents often bond more with a stable, reliable home community rather than variable local authorities that sometimes promise public investments in infrastructure or housing during local site visits and do not implementer them afterwards.

Sustainable office building in France

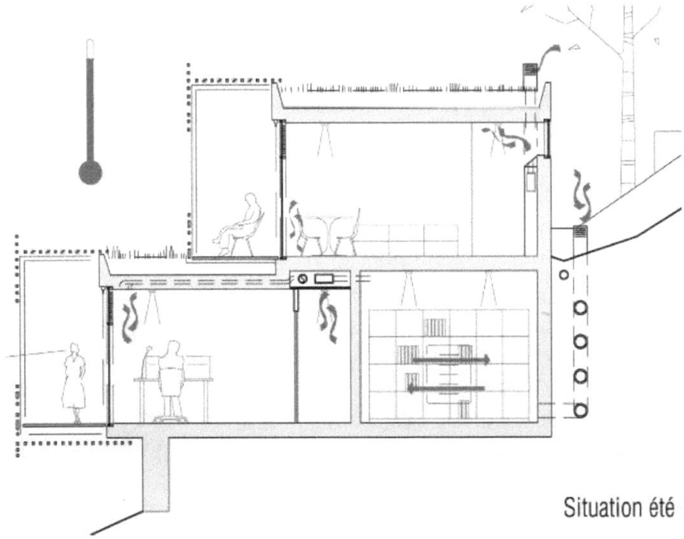

© rethink & Bernhard Lenz

For New Fayoum, an easy starting point would be a regular, reliable, and inexpensive public bus service linking New Fayoum to Old Fayoum and other interesting regional attractions such as Lake Karoum. A similarly helpful step would be a mobile grocery shop: selling fresh bread, vegetables, fruits and other groceries from a truck on a regular daily schedule, parked at a central spot.[9] Other services could include a school bus, a mobile health service, a mobile movie theatre, and public concerts. Soccer teams could be created in the new urban communities and competitions organised.

Similar mobile services have been developed in Europe in rural areas, where population density is too low to allow permanent facilities. In new towns, the context is different, but that model could be used for those services that are not economically feasible in the town's early stages. The public authorities would finance these temporary services while New Fayoum is in its low-population stage. Later, these services could be passed to the private sector. Most importantly, the first families to settle would be supported by a reliable service infrastructure. They'd

9 In Germany, the company Heiko (http://www.heiko.info) delivers with mobile grocery shops food products to about 1,600 villages in the rural regions of Cologne and Trier.

have opportunities for social interaction; these mobile facilities would also create a first round of jobs.

First German "Passivhaus"

©Passivhaus Institut

IV. How and For Whom Do We Build?

Sustainability is a global issue and a particularly important point for Egypt. In a country where 95% of the surface is desert, on-going urbanisation and climate change might have a strong negative impact on the fragile existing fertile area along the Nile River. The only arable regions in Egypt are the green floodplains that line the Nile River. This region, however, is also where most new urban development and population expansion is occurring. Today, Egypt imports a significant amount of its wheat, corn, and other staples, and spends billions a year in food subsidies. How will the country handle these supply concerns when its population rises above 100 million inhabitants while average temperatures are rising?

To begin, the most sustainable building is the one you do not build. In every country and every case, the construction process and the deconstruction process use lots of energy and materials. Construction affects local climate by replacing vegetation and repurposing agricultural territory, among other results. So an empty building is really unsustainable.

To make sure that any building is well planned and utilized, the future population should be involved as soon as possible in the design and construction process. Involvement does not mean that all decisions should be laid in the hands of the future inhabitants, as far as they can be identified. However, it does mean that they should ask questions and their concerns should be heard. In the case of New Fayoum, some of the new residents might come from nearby Old Fayoum, a town that continues to expand and built on the fertile agriculture land of the Fayoum Oasis. Interviews with the inhabitants about Old Fayoum's urban and economic structure and its social networks could result in better first-hand, local perspective. Such interviews would also be a good communication opportunity for New Fayoum and the need to protect the fertile land from construction.

Trying to relocate rapidly the population of Old Fayoum to New Fayoum would destroy the existing social network that people have created and rely on. Instead it is important to explain and discuss the need to slow and potentially stop the trend of building on fertile land while the community is still growing. Farmers normally know and understand the value of fertile land. The goal of preserving this land might be reached if New Fayoum becomes a real alternative to Old Fayoum, a livable and dynamic town. Bringing Old Fayoum residents into planning discussions can help build strategies and foster an urban/rural understanding for stabilising the border between town and Oasis.

Generally buildings should be adapted and planned for their geographical, climatic, cultural, and social contexts. The context of a building should be examined on a larger scale, such as the evolution of the region's hot, dry desert climate, but also on a local scale, such as considering how neighbourhood position affects shade thrown from other buildings or the presence of green areas. Passive energy-saving methods should be preferred to active systems, and the active system choice should be adapted to the technical knowledge of the users, since they will have to maintain it. Active systems that are too complex or expensive will not be maintained and will fail in short or long term. Again, involving the population in the process becomes an important step. Who do we build for? What are their needs and what are their means?

The first passive house, a low-energy house for a cold continental climate, was built in Germany in 1991 by Dr Wolfgang Feist.[10] The basic principles (good insulation, solar energy, and the like) had been developed by Wolfgang Feist and Bo Adamson in the 1980s. It began as an experimental prototype for interested test families, and then succeeded in Germany in the 1990s.

10 See the Passivhaus Insitut (http://www.passiv.de).

The idea began to spread through Europe in the 2000s. Today, passive buildings are being built in France, and the concepts have been adapted to schools, office buildings, and other building types. Research is under way to further adapt the concept to southern Europe, where the energy use for cooling in summer is often higher than for heating in winter.

These pilot projects and prototypes, such as the stabilized earth building in the courtyard of the Housing and Building National Research Centre in Cairo, are good ways to work out affordable, low-tech solutions and to communicate and share the quality of new sustainable buildings. The prototypes should be sufficiently tested before implementing them on a larger scale, since the early examples must be absolutely successful in order to gain the confidence of the general population. Sustainable solutions will not have a big impact if they stay in the hands of experts, or are too complicated to be easily shared. Sustainability must involve the whole society.

V. The Role of Housing

Even if many countries, like Egypt, urgently need homes for its growing population, housing should no longer be thought of purely as provision of shelter to the population. Rapidly constructing apartment blocks often leads to refusal and empty buildings or dormitory towns. In the development of liveable towns, housing is a strong instrument in delivering resilient and sustainable urban communities. Instead of being polarized and often mono-functional, any future housing strategy will need to offer complete and graduated housing plans on different planning scales from the building over the local area to the entire town. Housing presents one of the major cornerstones of urban development strategies.

Instead of thinking of housing in terms of one land-use category among others, as indicated in the master plan of New Fayoum, but also visible in the master plan of the new town of Sidi Abdellah in Algeria, it must be seen in terms of specific, multifunctional, and deliverable projects. Housing strategies help ensure that the local economic and social networks have a future. Speculative housing developments, where little is achieved in terms of supporting the local economy, should be stopped, as they do not lead to sustainable urban communities. Future housing projects should encourage distinctive urban quarters and establish effective, work-based, and inclusive neighbourhoods, while supporting the development of a "learning city". Positioning housing simply as shelter will never lead to successful projects that attract new inhabitants.

The relationship between the housing mix that the new town is offering in terms of flat sizes, building quality, urban forms and the estimated future populations with its characteristics (family size, habits, budget) and the economic strategy and the jobs it will offer (agriculture, handcraft, industry, service) must be studied and continuously adapted to the changes of the project. When the job offer is adapted to the characteristics of the diversified housing offer, the new inhabitants might be able to work close to their homes and avoid time-consuming and unsustainable daily transports.

Traffic in the Cairo city centre

© author 2012

VI. The Need for a General Consensus on Livable Cities

To succeed in the long terms, societies must agree to build better, more livable, more energy-efficient, and more environmentally conscious construction projects. Reaching this goal might include renouncing some elements that are part of our everyday life and are estimated today as important for personal fulfillment. For example, Germany's success in developing renewable energy, Austria's great development of contemporary wooden houses, and Sweden's high rate of waste recycling are based on the general consensus of their citizens. Their populations believe in those choices. Such goals depend on spreading knowledge and convincing people. Governments can promote sustainability issues, but they cannot create a general movement without popular support.

These new approaches require new forms of governance for the planning and construction of the new towns and later for their management. An inter-ministerial governance, with horizontal and collaborative structures, setting up a multidisciplinary steering unit for each new town with all the necessary skills and personnel (design, production, mediation, communication, and management control). These steering units, based and working in the new town, have the objective of pooling resources in order to achieve a shared project on all the different scales, ensuring the consistency and quality of the projects and control the proposals and the work performed by the private and public actors. They must have the authority and independence to coordinate and conduct the dialogue with all stakeholders (public, private). A scientific committee of outside experts from diverse backgrounds can meet in regular intervals to discuss progress, assess the quality of ongoing projects and studies and provide an outside view. The steering unit can then set up a subsidiary to manage the new town. A public management, which will address the needs of the new town and manage in a few years, depending on the success of the new town, the spontaneous constructions or organize the possible densification of parts of the new city. So that at a medium and long term an independent authority is created for each new town.

Project for the renovation of the historic quarter Bas-as-Salam in Damask in Syria

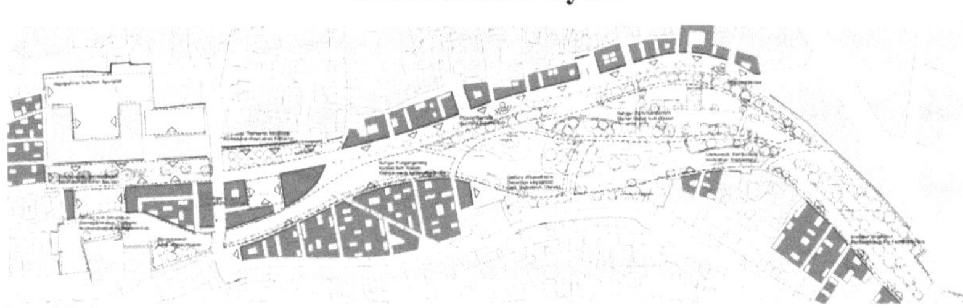

© rethink

Regulations play their roles in reaching the common goal, and might be necessary to transform the general idea of the urban form and the economic, social, and environmental frameworks in a more detailed and tangible understanding. But they should be developed carefully, as they have the nasty habit to multiply easily. Once they exist, the job of the public authorities is not finished, as regulations do not automatically create livable cities or good architecture. If they are not explained, not understood, or not monitored and enforced, they can be ignored or bypassed.

The different stakeholders should focus on developing a common and shared vision of life in the new town of New Fayoum. This vision should be worked out with the local actors and populations, and compared with national and international experiences. It does not have to be complete or perfect from the start, as it can be constructed in an iterative way, and can initially focus on some basic points. Once a common approach is found, it should be clearly communicated, to ensure that a majority of people approve and support the shared vision of the project. Workshops, architectural and urban competitions, debates, and visits can be good tools to reach this goal, and to get valuable feedback to integrate in the overall strategy.

If the political liberalization in Egypt continues, from the national to the local level, the messages given to inhabitants will matter. In that case, all that will be said and done in concrete actions should contribute to the main message – any intervention, any image, any reaction would count. The decision makers and urban planners will need to have the courage of their convictions.

Bibliography

Barthel, P.A., "Relire le Grand Caire au miroir de la densité", in *Confluences Méditerranée*, 75:4, 2010, pp. 121-135.

Pliez, O., *Les cités du désert, Des villes sahariennes aux saharatowns*, Presses universitaires du Mirail, 2011.

Subra, P., *Géopolitique de l'aménagement du territoire*, Paris, Armand Colin, 2007.

Sueur, J.-P. – Sénateur, *Villes du futur, futur des villes: Tome 1: enjeux*, Editions Sénat, 2011.

Sueur, J.-P. – Sénateur, *Villes du futur, futur des villes: Tome 2: analyses*, Editions Sénat, 2011.

Warnier, B., *Cergy-Pontoise du projet à la réalité: Atlas commenté*, Sprimont, Mardaga, 2004.

L'urgence permanente – The permanent emergency, Compte-rendu des Rencontres internationales de Venise et Marseille (juin-octobre 2000), Éditeur NEJMP, 2002.

The State of African Cities 2010, Governance, Inequality and Urban Land Markets, UN-Habitat, 2010.

Improving informal areas in greater Cairo, The cases of Ezzbet Al Nasr & Dayer El Nahia, Results of a case study project, TU Berlin, Urban Management Studies, 2010.

Heiko, http://www.heiko.info.

Passivhaus Institut, http://www.passiv.de.

Young Cities Project, http://www.youngcities.org.

Transparency International, http://www.transparency.org.

United Nations, department of economic and social affairs, population division, http://www.un.org/esa/population/unpop.htm.

The Principles of Sustainable Development

Fostering the Sustainable Planning Approach in New City Projects

Lessons Learned and Proposals (Morocco and Egypt)

Pierre-Arnaud BARTHEL

Associate Professor at the Institut Français d'Urbanisme (Paris-Est University) and researcher at LATTS (Laboratory for Technics, Territories and Societies – CNRS 8134)

Despite the new city tool being more controversial in Europe in the wake of an often mixed assessment of experiences, there are justifiable reasons to advocate for new cities in Arab countries and worldwide. Strong demographic growth, unregulated land subdivisions and occupations and the battle against urban spread aimed at preserving resources (farmland along the Nile, or Moroccan and Algerian valley floors) are just some of the factors behind the creation of new urban areas.

Furthermore, we cannot simply say that the new cities have no relevance; we should instead examine a number of failures by using the Egyptian experience as a reference country. We can then revisit a number of recent experiences in Morocco and Egypt which indicate the beginnings of a more quality and environment-oriented philosophy. Following this, a number of proposals can be put forward to developers to assist them in optimising their sustainable town planning approach.

I. Lessons Learned in Egypt: the Beginnings of a New Approach to New Cities

A. On the Ground, an Unsustainable Model in Egypt

There is an abundance of scientific literature on Egypt's new cities, covering their beginnings and their changes over 30 years.[1] Of all the

[1] Depaule, J.-C. and G. El-Kadi, "New settlements. Une réponse à la surpopulation?", in *Modes d'urbanisation en Égypte*, No. 1, 1990, pp. 187-198; El-Kadi, G., "Trente

countries in the Arab world, Egypt stands out from the rest for its return to the policy of new cities. Authorities in Sadat have embraced the mechanism, with a national program having taken shape. The program involves 22 new cities and began back in 1977 (10th of Ramadan was the first, with New Fayoum the latest launched in 2000). The General Organization for Physical Planning (GOPP), in charge of central planning under the Ministry of Housing, Utilities and Urban Communities, distinguishes between three types: satellite cities (6th of October, Al Obour), twin cities (New Damietta, New Asiut) and independent cities (Sadat, 10th of Ramadan). Furthermore, 24 others were earmarked in the mid-1990s for construction by 2030. The New Urban Communities Authority (NUCA) set up in 1979 under the aforementioned Ministry is the developer for this.

The NUCA is an extremely powerful public body, with very centralized operations (having branches devoted to each of the 46 new cities already or soon to be launched and with no independent planning or management functions). It managed a sizeable budget for the 2000s in the region of 10 billion Egyptian pounds (EGP) per year (9.8 in 2007/2008)[2] with revenues from the sale of serviced land accounting for about half of this. This makes it the fifth largest economic player in the country, just behind the Suez Canal agency and oil agency.[3] The NUCA thus manages the sale of land, establishes the objectives of each new city (population level to reach, jobs, economic and fiscal measures, specialization of the programs, etc.), negotiates construction permits for developments, and handles urban management aspects, including cleanliness, parks and urban networks. The very size of new city projects varies case-by-case. It is especially large in Cairo and even notable in the case of 6th of October whose perimeter located to the west of the Nile is equivalent to that of the central zone. The process of extending

ans de planification urbaine au Caire", in *Revue Tiers-Monde*, t. 31, No. 121, P.U.F., 1990, pp. 185-207; Denis, E., "La financiarisation du foncier observée à partir des métropoles égyptiennes et indiennes", in *Revue Tiers Monde*, 206, 2011, pp. 139-158; Denis, E., "From the Walled City to the Walled Community: Spectres of Risk, Enclaves of Affluence in Neo-liberal Cairo", in D. Singerman and P. Amar (eds.), Cairo Cosmopolitan: Politics, Culture, and Urban Space in the New Middle East, Cairo – New York: The American University in Cairo Press, 2006, pp. 41-71; Florin, B., "Vivre en parallèle? L'évolution des villes nouvelles du Grand Caire", in *Les Annales de la Recherche Urbaine*, 2005, pp. 97-105; Jossifort, S., "L'aventure des villes nouvelles. Vingt ans après, bilan et débats", in *Géographies de l'Egypte 2*, 23, 1995, pp. 169-194; Bayoumi W., "Greater Cairo New Towns 1977-2007: Obstacles and Mechanisms in the Voluntary Resettlement Process", Unpublished PhD Dissertation, University of Manchester, UK, 2009.

2 Sims, D., *Understanding Cairo. The Logic of a City Out of Control*, AUC Press, 2010.

3 *Ibid.*

the borders of certain new cities in Greater Cairo has been stepped up on several occasions, increasing the volume of space occupied.[4] The planning for housing units follows a strict approach of supply quantity while consuming large amounts of space with the general policy being that of "anti-density" designed to contrast with the central city.[5] And the trend over the years has been to move from programs for hundreds of units to developments of several thousand, often of one or two typologies maximum. Aside from private luxury developments, the landscape is interspersed with large low-cost housing developments, the latest being connected with the National Housing Program (NHP)[6] which was launched in 2005. A social mix, NUCA management claims, instead of a compartmentalized landscape, made up of socially homogenous pockets separated by over-sized highways. And the population growth goals for the planned areas are way up there. The authorities were expecting 5 million inhabitants in the new cities of Cairo in 2006, while the census indicates only 1.2 million. In 2006, 6th of October and Sheikh Zayed had 220,441 inhabitants (157,135 in the first and 29,499 in the second) while back in 1996 there were only 35,353. In 2011, some 700,000 inhabitants were living in this same area according to the latest studies[7] – a figure which seems plausible and shows an acceleration.

The history of new cities in Greater Cairo is well known; we need only look at the vast amount of research conducted on them. The 1980s were devoted to social projects for public housing and also the construction of industrial zones, notably 6th of October and 10th of Ramadan, with the NUCA itself, in its reports, quick to highlight a certain degree of success here.[8] The idea then was to attract, on the one hand, the working class, by constructing housing blocks at low cost and subsidized by the State, and on the other, private investors, through sales of

4 At the end of 2009, 80% of the 264 km^2 of al-Qahira al-Gadida (New Cairo) open to private development to the east of Cairo was sold. And new land expansions of more than 100 km^2 were, in the final years leading up to the "revolution", put up for auction by the NUCA.

5 Sims, D., *Understanding Cairo. The Logic of a City Out of Control*, AUC Press, 2010.

6 Sims also points out that all the NHP programmes require housing of the same size (63 m^2) – which tends to reduce the opportunities for plurality both in the neighbourhood and within the block and building. Furthermore, this expert is very familiar with Egypt's housing policies and shows that some State programmes deliberately aim at separating social groups. Sims, D., *op. cit.*

7 ArchPlan/AS&P, "The general strategic plan for 6th of October and Sheikh Zayed cities as one city", Masterplan, Projects and Transportation Strategy, GOPP-NUCA, 2009, 3 volumes.

8 Bayoumi W., "Greater Cairo New Towns 1977-2007: Obstacles and Mechanisms in the Voluntary Resettlement Process", Unpublished PhD Dissertation, University of Manchester, UK, 2009.

serviced land, even though at the time the private real estate development sector was still in its infancy. The zoning of urban functions was designed to bring about coherence; and there were reasons for the absence of public transport connecting with the central zone: the goal was to create an independent city, fed by the overflow of people from the capital which had become "unlivable", and offering a quick alternative to the unofficial zones with the aim of housing large numbers.[9] With Minister Mohammed Ibrahim Suleiman,[10] in office from 1993 to 2006, the period 1990-2000 was marked by a dramatic shift in the approaches to urban production, becoming increasingly governed by the financing of land and the arrival of national and foreign private developers targeting middle to luxury range real estate products. And the idea put forward by the Ministry of Housing, Utilities and Urban Communities and successive governments to justify putting land up for sale was that the arrival of well-off households would stimulate the arrival of lower classes and thus employment connected with construction and services for the wealthy residents.

Eric Denis points out that, over a period of 15 years (1995-2010), 1,200 km^2 were earmarked for private real estate development, representing an area equivalent to more than 2.5 times the agglomeration of Cairo which took 100 years to grow from 35 to 480 km^2 in 2000.[11] Naturally, the area constructed did not cover the surface area sold. Between 1995 and 2007, Cairo expanded with 110 km^2 of new constructions on desert plains and 55 km^2 on valley farmland. But now all the vacant land located within a 50 km radius of Cairo is for the most part no longer in the hands of the State. Cairo's new cities have therefore been playing fields for private investors, with investment opportunities in the center few and far between (rents blocked, public real estate assets frozen). Eric Denis also clearly demonstrates the extent to which the privatization of public land holdings, such as the illegal conversion of farmland for urban development, is mutually dependent on growth in Egypt's financial sector: credit deregulation, reopening of stock markets (in 1992) and laws favoring private investment.[12] The presence of prop-

---

9 This argument is well known in France where there are always defenders of the cause. Jean-Pierre Lacaze, honorary structural engineer, only recently wrote: "to provide a realistic answer to the housing shortage, there is no other option but to relaunch land development through public initiatives in sufficient quantity (3,000 to 10,000 houses)", Lacaze, J.-P., "De nouvelles villes nouvelles?", in *Dictionnaire de l'aménagement du territoire*, 2009, p. 70.

10 He was arrested on 6 April 2011 following allegations of corruption and illegally selling public land. His successor Ahmed Maghrabi was also arrested.

11 Denis, E., "La financiarisation du foncier observée à partir des métropoles égyptiennes et indiennes", in *Revue Tiers Monde*, 206, 2011, pp. 139-158.

12 *Ibid.*

erty companies is growing year by year on the EGX30, Egypt's stock market index, a reflection of the accelerating structured financing of land and real estate. The NUCA itself is building up its capital more and more on the markets in order to develop infrastructure: two loans were set up and secured using its land holdings with a guarantee from the Ministry of Finance in 2009. An auction system in 2005-2006 marked a major departure in what up until then had been rather opaque transfers of land by the NUCA.[13] Prices skyrocketed: in 2007, a single sale to Arab investors from the Gulf brought in some USD 3.1 billion, equivalent to 10% of the State's budget or 117 times the amount of property tax revenues collected.[14] The land sold comes with building rights that are still as minimal as ever in terms of surface areas (an average "open space ratio" of 0.5), while the terms and conditions applicable to developers remain as loose as ever – at the most, requirements for part of the development (less than 10% in general) to be low-cost housing as stipulated by the NHP.[15] The State sells to the highest bidder while certain Egyptian developers denounce the mad rush forward of a system of structured financing entirely focused on well-off households targeted with real estate products put on the market (Table 1). The years leading up to the international crisis were marked by heavy land and real estate price inflation linked to a maturing private development market in Egypt alongside large amounts of investments stemming from development holdings in the Gulf countries.

---

13 The most famous transaction was for Dreamland, a complex of gated communities with a golf course, 5-star hotel and leisure park, opened in 1991. 90 km² of exclusive land six kilometres from the Pyramids were transferred at a third of a piastre per m² to the Bahgat group. Florin, B., "Faire la ville hors la ville ou l'extraterritorialité des compounds, quartiers fermés du Grand Caire", *Regards Sociologiques*, Strasbourg, 2004, pp. 93-104.

14 Denis, E., *op. cit.*

15 The Madinaty project illustrates this new generation of more socially mixed residential developments. In exchange for a very attractive land selling price (no less than 34 km² of land in the New Cairo area!) the group has to earmark for sale 7% of its development for housing targeting low-income households. In autumn 2010, the transaction was the subject of a breach of contract; the State having sold the land to the investor at a price well below market rates. A major scandal followed just when the first owners were moving in.

Table 1. Winners of land auctions in Cairo's New Cities on 12 and 13 May 2007

EGP = Egyptian pounds (with EGP 8 = approx. EUR 1); acre = 4,200 m²

Name of developer	Location	Surface area in acres	Purchase price EGP/m²	Total value EGP
Damac (UAE)	New Cairo	1,500	752.5	4,740,750,000
Barwa (Qatar)	New Cairo	1,980	733.5	6,099,786,000
Ro'ya (Egypt)	New Cairo	460	1,288	2,488,416,000
Lake Side (Egypt-Libya)	New Cairo	80	4,005	1,345,680,000
Al Hokair (Saudi Arabia)	New Cairo	210	1,302	1,148,364,000
Mesha'l (Saudi Arabia)	6th of October Sheikh Zayed	410	852	1,467,144,000
Total		4,640	887.2 (ave)	17,290,140,000

SODIC and Global Research, 2009

Given the low population numbers living in the new cities in 2011, the reports all highlight the extremely large amount of public money spent on building drinking water supply stations, water purification stations and electric wiring for the new cities.[16] Since 1979, some 5 billion dollars have reportedly been spent on infrastructure for all of Egypt's new cities. This public money therefore also benefits the buyers of private villa[17] developments. The property tax exemption in the new cities favours this hoarding and no tax serves to restrict the vacancy rate which is almost 60% in Greater Cairo's new cities. Finally, the tendency prior to the 2011 "revolution" was to have at most the infrastructure paid by private developers.[18]

B. (Re)thinking New Cities: 6th of October and New City Re-planning

The territorial planning of new cities remains a sensitive subject. Putting together a new living space on large expanses of land (often several hundred or even several thousand hectares) is a challenge, and developers of new cities in Morocco and Egypt find it difficult to avoid the

---

16 World Bank (David Sims), *Arab Republic of Egypt: Urban sector*, 2008, 2 volumes; IDSC, "Sustainable cities in Egypt. Learning from Experience: Potentials and Preconditions for New Cities in Desert Areas", 2009.

17 A golf course villa bought for USD 250 000 in 1996, would have sold for at least three times that price in 2010.

18 Sims, D., *op. cit.*

temptation to produce fully drawn up master-plans, even though the phasing is such that it will be decades before the development is completed. However, the planning method is evolving. Workshops were organised by CDG Development for the new city of Zenata, bringing together a panel of high-level experts in 2006, along similar lines to the "New Medina" project which provides opportunities for improving the designs for the new cities of Chrafte near Tangier, Sidi Abdallah in Algeria and New Fayoum City in Egypt. Egypt, of all the countries in the region, is the country with the most experience when it comes to new cities. The way their development is being approached is changing dramatically.

At the end of the 2000s, a revamp in the planning of the country's new cities represented a shift in the urban development of these desert areas. Launched at the end of 2007 by the Egyptian Ministry of Housing, Utilities and Urban Communities, 6th of October is one of the test cities used as a basis for developing a strategic plan for the future.

Figure 1. New strategy for the future of the "New City" of 6th of October in Cairo

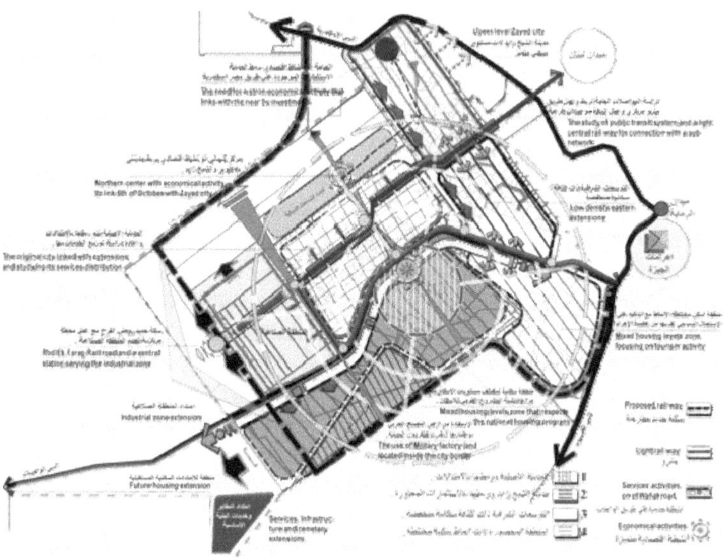

ArchPlan, 2009

In partnership with the German firm Albert Speer and Partners (AS&P), Ayman Ashour, lecturer in architecture and town planning at Ain Shams University, won the call for tenders for a sponsored study, together with his research and studies department called ArchPlan:

> Our vision is to develop a sustainable community for more than three million inhabitants to the west of Cairo founded on, both independence in its capacity to offer all functions possible and integration into the greater metropolitan region as a whole. In this business, the issue is two-fold: upgrade the existing elements of the new city and think ahead to future expansion plans.[19]

The strategy put forward by the consultants represents an original approach designed to address, from a "sustainable development" standpoint, the lack of centrality, designated public transport lanes, social mix, facilities and "nature" (within speech marks given the desert environment). The discourse is original; a valiant attempt at urban "remedial" work, rethinking the approach which until now has produced mixed results, even though some Western concepts appear to be blindly applied to Cairo: is green infrastructure the most appropriate tool in the middle of the desert?[20] Or will a tramway or bus way even be profitable enough, and therefore financially viable, given the low population densities? It remains to be seen to what extent the Ministry takes up this new approach proposed by the consultants and, with this as a starting point, if at policy level this new strategy will be "supported" at ministerial and governmental level. Difficult to say: discussions held with management are somewhat contrasting, between continuing the routine of this mega-administration and showing a real will to evolve.

Finally, the lack of substance in the environmental assessment of 6th of October is patent, including in the ArchPlan study. A gap in the thinking it appears... On this subject, few, possibly no studies on the cost of urban expansion (in terms of infrastructure and urban services) or on resource consumption (water, land, electricity) which very often takes place at the expense of neighborhoods in the central zone. During summer 2009, water distribution problems indeed occurred in the Madinat Nasr neighborhood due to water consumption in New Cairo, an area located just to the east of Madinat Nasr. A new pumping station for New Cairo's urban extensions has since been built, however the balance appears to be somewhat delicate, all the more so given that water poverty is a serious concern and that a hundred or so gated communities were under construction or completed in 2010 in Greater Cairo, all equipped with pools and golf courses.[21] The NUCA fixed the water price at EGP 1.05/m^2 for the new cities, triple the price paid in the rest of

---

19 Barthel, P.-A. and S. Monqid, 2011: interview with Ayman Ashour published, p. 57.

20 The consultant nevertheless has an interesting thought on nature in the city with the proposal for a large 140 ha park with one part green, the other drier.

21 Ibrahim, A. and D. Shaw, "Egyptian Compact-City Veracity: The Fallacy and Credibility of Mixed Use and High Density for a Low Carbon City", 45th ISOCARP Congress, 2009.

Egypt. Nevertheless, this price remains well below the production and related distribution costs.

Therefore, the new city tool is not being questioned as such in the different studies and the official line of the Egyptian Ministry and different governments. Indeed, it remains relevant in the eyes of those involved. The "Egypt 2030" vision follows the national policy of completing new cities and building new ones all located in the desert. The issue remains to be moving away from the Western model and identifying strategies for more tailored development in Egypt.

C. Private Developers: Launching Pilot Experiments

Private developers are also beginning to change the way they build new cities in the country. From a long-term perspective, the innovations target two types of projects: either environmental projects whose extra cost is covered by the solvency of the households that are targeted (i.e. well-off) and that are familiar with the latest trends in "eco-neighborhoods" in countries abroad, whether visited or not; or low-cost housing projects going back to local traditional architecture with the aim of developing a new construction sector there. In the second case, the size of the operations and the economies of scale that result need to absorb the extra costs.

Figure 2. Eastown in Cairo

SODIC-Solidere, 2010

Carried out by SODIC (Egypt) in partnership with Solidere (Lebanon), the Eastown project in New Cairo City and its twin Westown located in 6th of October reflect the new trend (Table 2). The entire "package" of the eco-neighborhood can be found here: bioclimatic town planning and architecture, networked air conditioning system, waste-water (grey and brown) treatment for use in watering, large spaces set aside for slow traffic (pedestrians and cyclists), compact nature of the operation and use of Medina-style traditional town planning for one of the project's sectors. Town planning that is very international yet at the same time tailored to local life permeates this dual operation where we find cycle paths and large, shaded common areas at the foot of a high-rise business center and residential programs.[22] We could also cite Moroccan examples like Sindibad Gardens in Casablanca (developer: Alliances) or Casa-Anfa led by CDG Développement. The Haram City project in Cairo is an example of the second trend. Developed by Orascom Housing Communities, this project adopted Nubian architecture inspired by that of the Egyptian architect Hassan Fathy with the aim of accommodating low or middle-income households (Table 2). Between the two trends, some developers are focusing efforts on environmental innovations, even though the goals may often remain very modest and even though sustainability may sometimes be put into the same category as greening public spaces, which need to be extensive, despite being in the middle of the desert!

Table 2. Winners of land auctions in Cairo's New Cities on 12 and 13 May 2007

Project	Developer	Location	Project concept	Sustainable Urban Development innovations
Madinaty	Talaat Mustapha (Egypt)	New Cairo (eastern suburbs)	Integrated city with a small area set aside for housing for those on low incomes	Solar-powered air conditioning
Cairo Festival City	Al Futtaim (Emirates)	New Cairo (on the periphery edge)	Community for middle and upper classes and business/shopping hub (Ikea, etc.)	Plant for treating water for re-use in green spaces; networked aircondi-tioning system to save energy
Haram City	Orascom (Egypt)	6th of October	Integrated community designed for social housing	Passive vernacular architecture; horizon-tal density

22 Barthel, P.-A., "Rethinking "New Cities" of Cairo: Challenges to Reverse Unsustainable Development", Égypte/Monde arabe, Urban Sustainability in Cairo: a Provocative Question?, 2011, http://ema.revues.org/index2966.html.

Westown/ Eastown	SODIC (Egypt) + Solidere (Lebanon)	New Cairo and 6th of October	Open city centres	Bioclimatic town planning; vegetation species suitable for dry climates; "slow" mobility; passive vernacular architecture

Interviews with professionals, conducted by M. Curtelin, C. Frais and P.-A. Barthel (2010), and project documentation

II. Sustainable New Cities: Strategic Principles for their Planning in the Arab World

Drawing on these lessons learned, the proposals formulated here make a distinction between two levels of scale to be strongly linked together. In this second part, the scale of the new city is addressed. We will see three priority strategic principles for thinking about the new cities from a sustainable town planning standpoint: firstly the increased importance of a quality and environment-oriented philosophy, secondly, the use of international labels and other levers, and thirdly, the social dimension of new city projects. Finally, in the third part, the methods used to link with the existing city or the agglomeration scale will be clearly set out.

A. Planning: a New Quality and Environment-oriented Philosophy

We have seen it in the case of Cairo's new city 6th of October, the urban ideology that prevailed in the past when it came to planning new cities is now taking a completely new direction. The objective is clearly to break away from the zoning approach and to think about putting together one or several hierarchical centralities linked with the centralities in the rest of the neighboring urban area. The idea is also to use the existing circumstances, including in the middle of the desert. In Egypt, the concern is really to think about the new city in relation to winds, sunshine, water scarcity and the need to take on a certain "rusticity" in the so-called unnaturally "green" areas. The idea is finally to put an end to hygienism: the developers are indeed responsible, to the extent possible, for working on collecting rainwater (when they exist) in the open air, reprocessing domestic wastewater for use in watering planted areas, etc. Toolboxes exist to help developers in the design-planning work. CDG Développement for example accepted in 2010 to experiment with an approach of the French Environment and Energy Management Agency (ADEME) called "Environmental Approach to Urban Planning"

(*AEU – Approche Environnementale de l'Urbanisme*) on an urban extension project in Rabat. This type of methodology therefore looks to really address planning from an integrated and inter-sector standpoint. Tekné, the private firm which was commissioned by the Agence du Grand Lyon, a partner in the experiment, therefore evaluated the studies and the development plan produced by the multidisciplinary team of Yves Lion, prize-winning town planner in France, and proposed improvements for "slow" traffic, housing/activity mix and parks by proposing a central mall unifying the future neighborhood.

Figure 3. AEU – Rabat Al Boustane

ADEME, 2010

The issue surrounding urban, architectural and environmental quality is a priority to address. The work can be done in ways that are different, but that can be combined together. The first task is to balance, at partnership level (between developer, operators and public authorities), requirement levels notably in terms of construction. Rules need to be formalized and serve as prerequisite conditions for land transfers to the operators planning to do the construction work. In Europe, terms and conditions are becoming increasingly detailed as regards materials to use, those to avoid, work on insulating buildings ("double skin", for example), the type of energy to encourage for home heating, the number of parking places planned for the development (increasingly restricted ratios per home if there is public transport), the possible pooling of these parking places and other services, etc. This quality also needs governance in place to implement it throughout the process until final delivery. Developers cannot simply pull out of the game as soon as the land is sold to operators, but must instead stay involved by following and managing the process of formalizing the architectural plans for each housing block. We can conclude by pointing out that it does not seem appropriate to impose requirements on absolutely everything, a real strategy must be devised in such a way that it reflects and balances out the "targets" we want to focus on (materials, energy, services, water, etc.). A strategy for urban and environmental quality can be produced as progressive and iterative. It benefits from being balanced and from commitments from the parties involved (city, town planning office, developers, entrepreneurs, service operators, etc.). These commitments can be formalized through charters co-signed by the stakeholders in the new city, or can lead to agreements, etc.

B. International Labels and other Innovative Tools: How These Can Be Used

Developers are beginning to use labels for environmental quality available on the international market. This is certification that builders therefore have to pay for and which is often required in Europe, even though this is an additional cost. The SODIC-Solidere consortium takes a free, case-by-case approach to applying the requirements of the North American label LEED NC/ND or BREEAM (British equivalent) to its dual Eastown/Westown project in Cairo. The phosphate company OCP in Morocco also wanted to have some signature developments LEED-labeled for its (new) city project in Benguerir (north of Marrakesh).

There again, in the region's developing countries, the recommendation is to not require all developers-builders to obtain labels, especially given the noticeable extra cost involved and the fact that the certification procedure makes the process of setting up operations more cumbersome. On the other hand, we can well imagine the town planner deciding to impose a few labeled pilot developments to serve as a demonstration of the new know-how there where the level of requirements will have been pushed higher than for more traditional developments. Certification seems to us to be compatible with and suitable for larger-scale public facilities, or the luxury hotel sector or even affluent residential areas (the clientele may then be receptive and want to buy). In the Benguerir green city project, the initial strategy of imposing the LEED label on all of the project's developments evolved to highlighting two pilot developments (Mohammed VI University and the "urban center" of the first part) for the purposes of setting the example.[23]

---

[23] Following a consultancy assignment I took part in July 2012 sponsored by the OCP, contracting authority for Benguerir's green city.

Sustainable new cities must also be thought of without labels or systematic recourse to eco-technologies especially if they are costly. In other words, there are avenues out there for innovative town planning which does not copy only the solutions in the north and which is well suited to countries in the region, such as Egypt and Morocco. Among these avenues, assisted self-construction is a viable avenue to explore for new cities. In Egypt, the "Ebni Beitak" program (Build your home yourself) follows this approach, however the assistance is not enough, with consulting architects needed to help and eventually approve projects for households with certain conditions covering building knowledge, materials used and subsidy mechanisms for financing additional costs. The same applies to low-cost housing where the quality is too often mediocre. A monitoring mechanism should be set up to support developers' projects. There could be pilot experiments on housing blocks, with the town planner setting these aside for eco-construction. We can well imagine housing and building technical centers (such as the Housing and Building Research Center) assisting the contracting authority in training, raising awareness among and advising diverse groups: young households, disadvantaged households, or middle classes. Sharing experiences and educating will be essential to success here given that eco-construction can be negatively perceived by households, synonymous with its traditional and even archaic building materials and know-how.

The idea is to achieve the best financial and institutional arrangement with the right engineering support for builders (households or developers). It is not a case of adding more, sometimes costly, eco-technology devices (solar panels, etc.), but rather of encouraging basic principles of sound architectural design.

C. The Social Dimension of Sustainable Planning for New Cities

Land transfers will benefit from having conditions imposed on them, not only in terms of environmental, urban and architectural requirements as we have just seen, but also in relation to planning that obliges the developers-builders selected to incorporate functional and social mix into their work. For the social dimension, this mix to be built on the land of new cities will act as a powerful lever for the project's sustainability.

Initial experiments with eco-neighborhoods in the region's countries sometimes focus only on beneficiaries in wealthier classes, indeed similar to certain pioneering experiments in northern Europe in the 1990s (Sweden and Denmark). SODIC and Solidere, developers of the Eastown/Westown eco-neighborhoods in Cairo, target classes A and B in their residential programs, offering classes at high level of environmental quality. We can understand the reasoning behind such a choice: the additional construction cost means that builders are obliged to target the most solvent categories. But while such a strategy may operate in Europe, it is doubtful that an "eco" middle class exists in the Arab countries, a class that still seems somewhat unwilling to restrict how much it uses its air conditioning and cars. Furthermore, what is the point of designing eco-neighborhoods for wealthier "eco" classes that do not yet exist?

The entire issue revolves around striking the right balance between supply and demand, and inventing the right business model: the supply can be made up of low-cost housing drawing on local know-how (maximizing natural ventilation, protecting against the sun, "traditional" materials) and repeated on a large scale in order to save on construction costs. Certain basic choices mean that well positioned housing, in terms of performance, can be put up for sale, yet remain within reach of middle class resources. In Egypt, the skills of residents and those of developers (many of which are illegal) are a source of education to be tapped into and developed further for programs involving housing jointly built by a developer, coordinating architect and local construction companies, which are used to working on high-density construction projects, yet can provide well developed pedestrian areas.

Low-tech town planning and architecture are therefore viable avenues for the future through reinvestments in local know-how (for example waste recycling, construction and architecture), and provide interesting initial lessons learned[24] as shown in for example the studies conducted on vernacular architecture, sources of low-pollutant construction material, and the renewed use of Arab-Muslim vocabulary for the urban design of projects. In Morocco, a guide on good practices for construction and housing has been commissioned by the technical department for housing at the Ministry, a positive sign of the desire to share successful experiences. This is also an interesting way of tapping into the "traditional" practices of residents (in the area of waste re-use, water management and building expertise) falling under sustainable development, but often not bearing this name.

Finally, the social project will be further enhanced if town planners do not destroy what already exists in their areas. The plan is therefore to draw on the resources of the local residents. Town planner Philippe Panerai advocated: "there are 3 douars in this area, they need to be integrated, they were there, and it is also an issue of sustainability,

---

[24] Notably for eco-tourism hotel projects, as in Siwa, Egypt or the rehabilitation of the historical town of Esna, again in Egypt.

everything there has something to offer, the douar can evolve, unofficial housing can be improved".[25]

This composition with what already exists calls for breaking away from "starting from scratch" town planning which is still too ingrained in the minds of decision-makers as soon as new city is mentioned. A sensible approach for the area means beginning with the land and "making do": if there is a plot of farmland (it can serve to guide the composition and preserve a memory of the site); if there is unofficial housing (it can be brought up to standard and blend in with the subsequent elements of the program); if there are sources and resources to tap into (wind, sun to be protected from or exploited, natural cycles, natural landscapes and their views), etc.

III. From Eco-Project to Eco-City: the Challenge of Change in Scale

The new city suffered in the 20th century from being viewed as an independent entity, disconnected from an urbanized environment. When located within an agglomeration, one strategy is to construct with the aim of going beyond the boundaries of the new city's perimeter and creating synergies between this new city and the rest of the built-up area.

In Benguerir, an agglomeration of 60,000 residents located to the north of Marrakesh on the road leading to Casablanca, a "green city" project was inaugurated by the King of Morocco himself in 2009 on a potential land area of 1,200 hectares (the existing city representing 2,600 ha and 600 ha already prepared for the development). The land earmarked for the project, we point this out because it represents an opportunity for the project, is located very close to the built-up areas. This morphological continuity between the first parts of the project (including Mohammed VI University) and the existing city makes the question of integrating the new city into the agglomeration extremely important.

A workshop of experts I took part in during July 2012 resulted in advice given to the phosphate company OCP, the green city's town planner. As regards integrating the green city, the idea was to say that the key challenge is to think about the OCP's green city project in terms of a future eco-city made up of the green city and the existing city. Indeed, positive concomitant changes were identified during the workshop. While, on one side, the green city is an experimental area-project with LEED showcase developments and a genuine concern for project quality ("green" strip, facilities, mix of functions), it appears that, on the other

25 Panerai, P., New Medina Meeting in Marseille, 9 March 2012.

side, there were major efforts underway to bring precarious neighborhoods up to standard (contracting authority Al Omrane), and to restore quality to public areas. Based on this, a common green city/existing city strategy needs to be developed through a joint effort of the parties involved (OCP, City of Benguerir, Governorate, Town Planning Office, etc.).

Figure 4. Benguerir green city project

OCP, 2010

Ideas have therefore been outlined as proposals for this strategy's approach. There are several levers that can be used: a transport plan common to both components so that at least one public transport line is put in place (Bus way recommended) and a related traffic plan; a common "green" network to extend the green zone of the OCP project in the existing city; a cross-strategy for housing (to avoid luxury housing in the green city project and low and middle range in the existing city); and a common water, waste and purification strategy (a new wastewater treatment plant is under construction). There was thus a need to explain to OCP the benefit of developing its concept of green city for all of Benguerir: for concerns over development and urban integration, and also for the complete success of the project itself. A simple concept was put forward: introduce a double centrality and a "crossing" from the existing station to the University of the Green City Project, with related work as detailed in Table 3.

Table 3. From the new city to the eco-city: strategy outline for Benguerir (Morocco) – proposals made by experts to OCP

	Centrality of the green city	Centrality of the existing city	University-Station crossing
Ad-vantages, potential	* International University * Urban centre * Pedestrian zone	* Existing station * Existing neigh-bourhoods * Fallow land (land available)	* Avenue de l'Université to the RN9: commercial hub * Existing facilities (secondary school, hospital), existing neighbourhoods
Actions	* University developments in parts * High-density urban centre with social mix * Pedestrian zone facilities available to everyone	* Station brought up to standard * Regeneration of the outskirts and fallow land (urban density)	* Construction of building facades * Change to urban boulevard * Introduce public transport with designat-ed lanes (e.g. Busway) * Facilities brought up to standard
OCP's role in the work	Leader, OCP: Financing and running the project	Leader: ONCF, local and national actors and support from OCP	Leader: Bus company to be set up (OCP in-volvement), local actors and the State

P.-A. Barthel for the ESSEC workshop, Urban Economics Chair, July 2012

Conclusion

To conclude, sustainable town planning aims to be present in all aspects of development operations: creating new neighborhoods, bringing unofficial neighborhoods up to standard, urban regeneration of historical features, renewing the industrial fabric, and of course new cities. For the latter, we believe a new attitude is emerging among some new city designers and decision-makers. Furthermore, the *New Medina* project draws on its case studies in order to help developers design new approaches and methodologies compared to previous projects. In Egypt, Algeria and Morocco the challenges are huge: the challenge of linking the new city in with public policies to take it out of its extra-territoriality situation, the challenge of integrating into the existing city, the challenge of effective governance which takes the developer out of his/her private domain and isolation, the challenge of the flexibility and mix of the urban composition being able to adapt over time, and the challenge of linking in with the issue of public transport in order to break away from the obsession with cars.

The new cities can remain an area of experimentation: well adapted eco-construction targeting certain groups in society, smaller-scale, sensitive and humanist town planning, innovation in certain fields (positive energy of some buildings, introduction of local industries in line with local assets), town planning that taps into local sources and resources (including the truly unloved desert).

Bibliography

ArchPlan/AS&P, "The general strategic plan for 6th of October and Sheikh Zayed cities as one city", Masterplan, Projects and Transportation Strategy, GOPP-NUCA, 2009, 3 volumes.

Barthel, P.-A., and L. Zaki (eds.), *Expérimenter la "ville durable" au sud de la Méditerranée*, Aube publications, "villes et territoires", ESSEC collection, 2011.

Barthel, P.-A. and S. Monqid, *Le Caire, réinventer la ville*, Autrement, Villes en mouvement collection, 2011.

Barthel P.-A., "Rethinking "New Cities" of Cairo: Challenges to Reverse Unsustainable Development", *Égypte/Monde arabe*, Urban Sustainability in Cairo: a Provocative Question?, 2011, http://ema.revues.org/index2966.html.

Barthel, P.-A., "First experiences of sustainable neighborhoods in Arab Mediterranean countries: reproducing the European model?", Lehmann S., Al Waer H., Al Qawasmi J. (ed.), *Sustainable Architecture and Urban Development*, Vol. 3, CSAAR Press, 2010, pp. 121-136.

Bayoumi, W., "Greater Cairo New Towns 1977-2007: Obstacles and Mechanisms in the Voluntary Resettlement Process", Unpublished PhD Dissertation, University of Manchester, UK, 2009.

Denis, E., "La financiarisation du foncier observée à partir des métropoles égyptiennes et indiennes", in *Revue Tiers Monde* 206, 2011, pp. 139-158.

Denis, E., "From the Walled City to the Walled Community: Spectres of Risk, Enclaves of Affluence in Neo-liberal Cairo", in D. Singerman and P. Amar (ed.), *Cairo Cosmopolitan: Politics, Culture, and Urban Space in the New Middle East*, Cairo – New York, The American University in Cairo Press, 2006, pp. 41-71.

Depaule, J.-C. and G. El-Kadi, "New settlements. Une réponse à la surpopulation?", in *Modes d'urbanisation en Égypte*, No. 1, 1990, pp. 187-198.

El-Kadi, G., "Trente ans de planification urbaine au Caire", *Revue Tiers-Monde*, t. 31, No. 121, P.U.F., 1990, pp. 185-207.

Florin, B., "Faire la ville hors la ville ou l'extraterritorialité des compounds, quartiers fermés du Grand Caire", *Regards Sociologiques*, Strasbourg, 2004, pp. 93-104.

Florin, B., "Vivre en parallèle? L'évolution des villes nouvelles du Grand Caire", in *Les Annales de la Recherche Urbaine*, 2005, pp. 97-105.

Ibrahim, A. and D. Shaw, "Egyptian Compact-City Veracity: The Fallacy and Credibility of Mixed Use and High Density for a Low Carbon City", 45th ISOCARP Congress, 2009.

IDSC, "Sustainable cities in Egypt. Learning from Experience: Potentials and Preconditions for New Cities in Desert Areas", 2009, 112 p.

Jossifort, S., "L'aventure des villes nouvelles. Vingt ans après, bilan et débats", in *Géographies de l'Egypte 2*, 23, 1995, pp. 169-194.

Lacaze, J.-P., "De nouvelles villes nouvelles?", Dictionnaire de l'aménagement du territoire, 2009, pp. 65-71.

Panerai, P., New Medina Meeting in Marseille, 9 March 2012.

Sims, D., *Understanding Cairo. The Logic of a City Out of Control*, AUC Press, 2010.

World Bank (David Sims), *Arab Republic of Egypt: Urban sector*, 2008, 2 volumes.

Between Orders and Appropriations, a Return to the Concept of Sustainable Urban Development in the Southern Mediterranean

Lauriane LAHERY

Project Officer at European New Towns and Pilot Cities Platform, ENTP

Introduction

In 2007, for the first time, the urban population accounts for more than half of the world's population. Henceforth, no development strategy can ignore the cities. Responsible for 75% of the emissions of greenhouse gases and the consumption of 75% of global energy,[1] cities are at the heart of global concerns, especially with regard to the fight against climate change.

The arrival of the notion of sustainable development in academic discourse and policy in the early 1990s brought forth urban issues in a global system, whereby local issues are equally considered as global threats, reinforcing the need for international cooperation. The concept of sustainability itself has multiple definitions and appears as a malleable instrument both inspiring and conducive to global consensus.[2] However, the term, once narrowed down to given areas such as cities, deserves to be studied in the context of cooperation projects.

Although supported by major international institutions, sustainable development or sustainable urban development in the case of urban projects carries with it the hope of changing the past developmentalists approaches evolving in a global capitalist model, and would "in theory"

1 Barbier, C. *et al.*, Introduction, in *Regards sur la Terre 2010. Villes, Changer de Trajectoire*, L'annuel du développement durable, Presses de Sciences Po, 2010.

2 Hours, B., "Le développement durable, instrument d'intégration globale", in Martin, J.Y. (ed.), *Développement durable: doctrines, pratiques, évaluations*, Paris, IRD Editions, 2002, p. 288.

enable change in the approaches taken towards development projects. Should we consider sustainability as a "must-have" of cooperation, attached to development projects, allowing a makeover of past actions, or as an innovative instrument?

Consisting of two spatial scales, from global to local and vice versa, it is the responsibility of the cooperating actors to juggle as they can with respect to these spatial requirements. Between ownership and international orders, how to build sustainable development projects in areas of urban cooperation?

It is by returning to the concept of sustainable development, focusing on the city, that we begin this analysis, before shifting towards a Mediterranean context where projects in Cairo such as project New Medina and interviews conducted in 2011[3] illustrate the use of sustainable development in urban cooperation.

I. Sustainable Urban Development, the Return to a Recent Concept

Development refers to a representation of the growth and improvement of living conditions. It makes its appearance on the global political scene in 1949, when the President of the United States Harry Truman delivered his speech on *the State of the Union*. In introducing the term underdevelopment, he opens the way to the concept of development.[4] Thus, there is no change to each individual understanding of it, but rather different progress reports modelled on this single Western model.

The institutionalization of development and its "globalization" started with the creation of the World Bank and the International Monetary Fund in 1946. These large financial organizations play a major role in what may be called the "orders" of development, indicating a procedure modelled on the West. NGOs and states, through development aid, are also actors using these arguments around the world, depending on their particular scope of actions.

The concept of development significantly evolved, especially with the advent of "sustainable development" in the late 1980s. It is defined by the 1987 Brundtland report as a "development that strives to meet the needs of the present without compromising the ability to meet those of future generations". We find this notion today in the field of urban

---

3 Interviews held in Cairo with international donors for the CEDEJ (Centre for economic, juridical and social studies) under Pierre-Arnaud Barthel monitoring, April to July 2011.

4 Tremblay, S., "Du concept de développement au concept de l'après-développement: trajectoire et repères théoriques", in Collection *Travaux et études en développement régional*, Université du Québec à Chicoutimi, December 1999.

development, with the advent of the term "sustainable city" or "sustainable urban development". How to adapt this concept, rather vague, to the city? It seems advisable to add, as Adriana Allen does in her article "Sustainable Cities or Sustainable Urbanization",[5] two additional pillars[6] supplementing the original definition: physical sustainability, an intervention that emphasizes sustainability for all urban structures as well as the efficiency of the built environment, and political sustainability that takes into account the quality of the system of governance that will guide the actions of the other actors in other regards including the participation of local civil society.

The last pillar is often called "political vision" and ensures a juridical framework, long-term financing, land control and coordination between the different administrative levels.[7] Sustainable development is thus meant to be qualitative and not only quantitative.

Urban development, however, requires significant financial resources and a coherent institutional framework. Multilateral and bilateral cooperation, defined as a policy of economic, financial and technical support from developed countries to developing ones,[8] contributes to the Official Development Assistance (ODA) since the 1970s in particular, when the fight against poverty, while confined to rural areas, extends to the city. Sustainable development as an instrument of global integration legitimates actions of international cooperation and integrates environmental considerations within urban issues.

II. The Diffusion of Sustainable Urban Development, New Adage of International Cooperation

The spread of sustainable development and sustainable urban development took place from the 1990s. Brought to their early days by UN agencies, these concepts gradually became essential to multilateral and bilateral cooperation. Presented as the only vehicle to changes that meet the requirements of development and environmental protection, sustainable development has given rise to new lines of conduct, under the influence of international organizations.

---

5 Allen, A., "Sustainable Cities or Sustainable Urbanization?", in *UCL's Journal of Sustainable Cities*, 2009.

6 The three traditional pillars are: social – economy – environment.

7 Shahriari, E., urban planner and architect, research assistant for Young Cities Project, Technical University of Berlin, quoted in the New Fayoum panel of experts' minutes, project New Medina, September 2012.

8 *Larousse dictionary*, 2011.

A. The City, a Growing Awareness in the International Scene

Urban development has not always been a priority for donors. The fight against poverty has long been confined to rural areas, even within the UN or the World Bank considered poverty in 1950-1960 "as a phenomenon of rural nature and the process of urbanization as a direct result of such poverty".[9] This element changes in the 1970s. The funds allocated to this sector gradually increase, reflecting a positive change for cites. This period is not trivial since it corresponds to the acceleration of urbanization and the increase in urban poverty. Robert McNamara, President of the World Bank from 1968 to 1981, stated in 1975: "If cities do not constructively deal with poverty, poverty may well become more destructive to cities".[10] In the 1980s, the decade of the debt crisis also represents a crisis for development: urban poverty increased, aid is diverted, state subsidies are cut and urban development results are disappointing.

B. The 1990s, the Promotion of Sustainable Urban Management and Participatory

The Rio Conference in 1992 formalized the promotion of sustainable models of human settlements and stresses the importance of sustainability in their management, especially in developing countries.[11] Instituting 'Local Agendas 21' in as planning tools for sustainable development, it marks an important breakthrough in urban management. These are based on the elaboration of a diagnostic and participatory action plan. The real impetus for the promotion of sustainable urban management remains the 1996 Habitat II Conference held in Istanbul. In total, 171 countries ratified the Habitat Agenda, which introduces inclusive and participatory urban development. The role of local actors in urban management – seen as multidisciplinary – is recognized.

The city then truly becomes the best place for sustainable development, which can be broken down into specific objectives for effective urban management, balanced and efficient.[12] Numerous methodologies dedicated to urban development have emerged. UN Habitat created

---

9 De Ponte, G., "L'évolution du discours sur la ville: des institutions multilatérales d'aide au développement", in *Revue internationale des sciences sociales*, No. 172, 2002, p. 232.

10 *Ibid.*, p. 233.

11 United Nations, *Agenda 21*, World Summit, Rio de Janeiro, chapter 7, paragraph 15 of Action 21, Rio declaration, 1992, accessed 4th of March 2011; http://www.un.org/french/ga/special/sids/agenda21/action7.htm.

12 Cavallier, G., "La nouvelle politique urbaine de la banque mondiale: un essai à transformer", in *Les annales de la recherche urbaine*, No. 86, 2000, p. 151.

"City Consultation", the World Bank created "City Development Strategy" (CDS), a tool for national and international communication used later by the Cities Alliance, a global partnership founded in 1999 whose aim is to fight against urban poverty and promote the role of cities for sustainable development. It is important to mention that new methodologies have emerged from Non-Governmental organizations (NGOs) and bilateral cooperation agencies. Some countries did not wait on international meetings to take initiatives: environmental concern emerged in Turkey in the 1970s as in Tunisia, and then in 1980s in Morocco and Algeria.[13]

III. The Sustainable City in the South of the Mediterranean: Foundations and Specificity

Texts after texts would emerge, although a vague and malleable definition of sustainable urban development remained. Do parties protect themselves from commitments that cannot meet their personal interests? If this concept is mainly conveyed by international development agencies and NGOs, is it not the reflection of a Western model of thinking that would make it difficult to adapt to the South? When disseminating standards and mechanisms to lead to sustainable urbanism in developing countries, international institutions seek to generalize "best practices", a kind of methodological packages indicating the path to follow and the agenda to adopt. Awareness of sustainability can be seen as a series of injunctions upon States to adopt more democratic management methods.[14] These recommendations may quickly appear as disconnected from the local political system and economic choices of countries. To focus our study in its proper geographical context, we should mention the specificity of urban development on the Southern Mediterranean region, observing the emergence of new methods of "healing" and rethinking the cities.

A. Urban Issues and Institutional Barriers

The cities of the south shore of the Mediterranean face many challenges that development policies are struggling to meet. Public policies have long focused on the production of housing, later accompanied by a sectorial policy of basic equipment in service, only rarely considering

---

13 Barthel, P.-A. and L. Zaki, *La Ville Durable au sud de la Méditerranée. Paroles de Chercheurs et de Professionnels*, Paris, Éditions de l'Aube, 2011, p. 5.

14 Navez-Bouchanine, F., "Le développement urbain durable: 'best practices' ou leurre méthodologique?", in *Espaces et sociétés*, No. 131, 2007, p. 101-116.

the plans and master plans when they exist.[15] This sectorial development of the city is often found unfit to meet the challenges of southern Mediterranean cities. In Tunisia, efforts to increase the weight of medium-sized cities have failed to slow the process of coastal development or concentration of activities in northeast areas. In Morocco, the weights of Casablanca or of territorial inequalities have not been reduced. Other countries have also been unable to contain the growth of major cities. Implemented plans have been overtaken by population growth and increasing informal housing, due to the lack of affordable housing.

Spatially spread and socially segregated, Cities are in crisis. The population's access to essential goods and services remains uneven and unemployment persists as well as pockets of poverty in both urban centers than in their periphery. In response, without harming the environment, a growing demand for services, equipment, housing or energy, promotion of new urban practices proved imperative. On top of these local issues, global concerns are equally important for the future development of cities. Due to its hot and dry climate, the southern shore of the Mediterranean will be faced with brutal impacts of global warming: There is an increased threat of desertification of arable land as well as the threat of rising waters to the coastlines and the Nile Delta. The end of oil reserves also affects those countries whose energy needs are growing each year.

Unmastered urban development and future environmental and energy challenges ahead are also vehicles that have accelerated the spread of sustainable development in the Arab world.[16]

The distribution of this referential came late compared to the north shore. This delay reflects an appropriation less obvious, less fluid. Lack of technical and financial resources could be an obstacle to the spread of sustainable development, just as the lack of real political leadership or the fear of slow economic growth that is sometimes at odds with environmental concerns were.[17]

The first related programs to sustainable urban development were engaged in the early 2000s, through the adoption of Local Agenda 21 plans or strategic development. While they were introduced in Europe by local organizations with the necessary financial and technical re-

---

15 Mediterranean Institute: *Des politiques urbaines et territoriales encore fortement centralisées et pilotées par l'État*, accessed 26th of July 2011 http://www.ins-med.org/2010/12/des-politiques-urbainesetterritoriales-encore-fortement-centralisees-et-pilotees-par-l%E2%80%99etat/.

16 Barthel, P.-A. and L. Zaki, *La Ville Durable au sud de la Méditerranée. Paroles de Chercheurs et de Professionnels*, Paris, Éditions de l'Aube, 2011.

17 *Ibid.*

sources, their application on the other is confronted with centralized management systems with more limited resources. The central government may resist, wishing to control the strategic choices of the country, or seeing sustainable development as disconnected from the economic choices and realities, from its priorities or as a requirement which may dampen economic growth. European cities have managed to form networks of communities engaged in urban sustainability in part through joint initiatives driven by the 1994 Aalborg Charter on Sustainable Cities. In the south, the commitment is made more complex by the absence of Arab supranational structures and the lack of protest, leading to a top down institutionalized sustainable development.[18]

If urban development is shown out of step with urban dynamics, then the fault is largely due to institutional structures, being faulty or non-existent and only having slight mastery of the land. The lack of coordination and common standards between departments or planning agencies with limited power creates a very sectorial approach of the city, without a common strategic vision and disconnected from local realities. "Ultimately, institutional fragmentation leads to urban catch up, rather than anticipatory planning, to the detriment of a comprehensive consideration of sustainable development".[19]

Will the centrality of power not be assessed in face of the revolts of 2011 in Arab countries, such as Egypt, Tunisia and Morocco which are currently undergoing political transition? These events have shown that there is a way other than the central government. They also showed that the current development model based on patronage and rent (energy, land) benefiting only the elite was no longer sustainable. The democratic wave in action, accompanied by pressure from multilateral agencies, may accelerate the process of decentralization and participation in the near future. This represents much hope for societies and for the future of urban planning.

B. Urban Sustainability on the South Shore, a Process that has Started

Without a binding framework, progress in terms of urban planning is quite reserved, but progress in this direction cannot be ignored. For the

---

18 Barthel, P.-A., "Premiers quartiers urbains "durables" dans les pays arabes: Enseignements et doutes sur une génération spontanée (Maroc, Tunisie et Égypte)", in Houpin, S. (dir.), *Identification de critères de mise à l'examen de projets urbains durables dans le cadre de l'Union pour la Méditerranée*, rapport final du Plan Bleu, 2011.

19 Mediterranean Institute: *Les limites des outils de planification urbaine et territoriale*, accessed 26 July 2011 http://www.ins-med.org/2010/12/les-limites-des-outils-de-planification-urbaine-et-territoriale/.

past ten years there has been a proliferation of urban initiatives on the south shore, which are modeled on the standard of sustainable development. We rediscover the *medina*, a place which masses the historical center of the Arab city. Trips and short journeys are made by walking, housing is compact, suitable climatic conditions and social relationships are often preserved.[20] This shows that sustainability does not solely lie in "high-tech" developments, and that Arab towns have a great potential. New neighborhoods described as "sustainable" are emerging, alternative energy projects multiply, taking advantage of the sun and wind as transport go green. Many public transport initiatives are indeed at work today in the Maghreb and the Middle East. To take a few examples, the third metro line in Cairo is under construction while the fourth is being planned, the Algiers metro was inaugurated on October 31 2011, and the tram on site Rabat-Salé in Morocco was inaugurated in July 2011.[21]

While these initiatives are not new, mobility is rediscovered through the prism of sustainable development.[22] The introduction of these standards in public transport projects gives them a new visibility. All these projects are designed, if successful, to be replicated to contribute to the definition and dissemination of "best practices" of sustainability across each country.[23] Initiatives, which would nowadays be labeled as sustainable, had already emerged in the southern Mediterranean countries. In Algeria in the 1980s, the Egyptian architect Hany El Miniawy launched the construction of green buildings in the village of El Oued in Algeria, 620 km south-east of Algiers, based on low-cost local technics and materials and training of local builders. Hany El Miniawy returned to Algeria a few years ago and found that many other buildings were built according to models he had created at the start: the spontaneous reproducibility is a guarantee of success.

---

20 Barthel, P.-A. and L. Zaki, *op. cit.*

21 http://www.enpi-info.eu/medportal/news/latest/25989/Le-Maroc-inaugure-le-tramway-Rabat-Sal%C3%A9-et-la-station-d%E2%80%99%C3%A9puration-d%E2%80%99Oujda,-deux-projets-financ%C3%A9s-par-la-BEI "IEPV" website in French, the European Neighbourhood and Partnership Instrument, accessed 26 July 2011.

22 Barthel, P.-A., *op. cit.*

23 Barthel, P.-A. et L. Zaki, *op. cit.*

School in El Oued, El-Miniawy Architects

The most famous neighborhood or rather "sustainable city" remains as Masdar City, launched in 2006 in Abu Dhabi. This city was created ex nihilo, as an environmentally friendly city because it does not emit carbon dioxide. This case represents very specific example, undertaken under quite exceptional circumstances because it attracts the attention of politicians, investors, and the financial resources deployed are spectacular.[24] Therefore, there is very little probability that this model can be replicated in other countries that do not possess the same means. Informal settlements have again become an important topic as there is an increasing awareness that more inclusive approaches to urban development should be incorporated.

The preoccupation regarding informal approaches is apparent in several Mediterranean countries, and reflects a willingness of authorities to revise their way of thinking about the city.

In Tunisia, starting in the 1980s, informal settlements have progressively been integrated into cities in order to fight the Islamist movement that was gaining popularity in the outskirts of the city.[25]

In Egypt, it is only 20 years later that the government has revised its way of interpreting the "illegal". In 1996, a decree provides that any construction without a permit or agricultural land will be severely punished. In 2005, the governorate of Cairo, faced with the failure of policies on informal approaches, invalidates the latter decree and an-

---

24 Houpin, S., *op. cit.*

25 Chabbi, M., "La révolution, malgré la réhabilitation des quartiers populaires", in *Villes en développement*, bulletin du Partenariat Français pour la Ville et les Territoires, No. 91, June 2012.

nounced a year later that all informal housing built before June 2006 would be allowed to be connected to the network, at least temporarily. It is now recognized that the prohibition in itself served no purpose, and that it is up to the state to offer alternatives.[26]

Alternatives are proving to be all the more necessary since the proliferation of informal settlements following the uprisings of 2011, favored by the disappearance of control agencies, have revealed the absence of social housing policy as well as the little diversification in the public offer of housing. If governments – still being formed – have remained silent, the international development agencies have on the contrary shown a growing interest in the informal areas. Without any partnership or governmental support, their recommendations are unlikely to be materialized on the ground. Popular committees, which were formed in Egypt during the revolution to ensure the security of neighborhoods, did not wait for actions "from above" to organize themselves and to meet the inhabitants' needs (waste collection, traffic management, safety, infrastructure rehabilitation…).[27]

Other examples of urban planning considered "sustainable" are often made by private developers touting "eco-neighborhoods". Sustainability is understood in its size firstly as "ecological" with the promotion of new architectural standards and greenery, the latter reflecting a marketing vision (golf, gardens and fountains). These new luxury neighborhoods are not truly accompanied by the politics of planning policy, which limits their "sustainability": social segregation, urban sprawl, land split, lack of public transport are all negative externalities that pose challenges the merits of these neighborhoods.

Even though there are still flaws at the institutional level and at the level of its implementation, sustainable development is now referred to by many players. We can insist that the reference to this concept can legitimize minor urban options that were not or were previously difficult to justify.[28] The sustainable urban development has opened a new forum for dialogue where the development model is revised so that it can anticipate, with a holistic and long-term approach, the future impact on territories, societies and economies.

---

26 Sims, D., *Understanding Cairo: The Logic of a City Out of Control*, AUC Press, 2011, pp. 129-130.

27 Sims, D., "Printemps arabe et quartiers informels au Caire", in *Villes en développement*, bulletin du Partenariat Français pour la Ville et les Territoires, No. 91, June 2012.

28 Clerc, V., "Développement urbain durable et quartiers informels à Damas: évolution des paradigmes et contradictions des réformes", in Barthel P.-A. and L. Zaki (dir.), *Expérimenter la "ville durable" au sud de la Méditerranée*, Paris, Éditions de l'Aube, collection Villes et territoires, 2011, pp. 175-199.

IV. Cooperation and Sustainable Development in Egypt, History of a Word Become Indispensable

A. Between Compromise, Stagnation and Ownership

Various cooperation projects try to promote a more integrated vision of urban development. Various studies – conducted in partnership with Egyptians or not – are not always used. The quantitative approach of the centralized Egyptian state leaves little room for anticipation and an integrated approach to the city.

Egyptian cities do not receive money directly from donors, and this could represent an obstacle in programs focusing on strengthening local capacities.

The successful implementation of sustainable urban projects is hampered by "institutional gaps" mentioned by several stakeholders interviewed in 2011, reflecting the lack of coordination and cooperation ("not hand shaking") between those who implement a particular action and those who manage and finance the project. Urban planning is greatly dependent on a budget that cannot or will not answer to its needs, and on an accountability that seems to be missing, hence the lack of concrete action. Are models proposed internationally not at a disconnection with reality? According to our many interviews conducted in 2011, models are built on local realities, and most of the time by a qualified local staff. If the message does not go through, the lack of collaboration/understanding between those who work for the sustainable management and those who have the power to implement seems to be the most suitable explanation for the lack of progress.

International actors, ranging from donors to NGOs, and through to networks of cities and experts, can be a tool for change or at for the emergence of different and innovative visions supporting local initiatives. However, support from authorities gives hope that the promoted initiatives will be maintained over time.

Would development that is driven by cooperation become more sustainable? If the situation is gradually changing in the right direction, it does not seem to change proportionally to the money spent and efforts made by the Egyptian actors internationally. W.J. Dorman[29] asks himself the same questions, wondering how the presence of donors, of at least 40 years, was unable to develop the institutional capacities of the state to solve the problems of the informal. For example, two explanations deserve to be mentioned here (for the period of 1974-1998), one origi-

29 Dorman, W.J., *The Politics of Neglect: the Egyptian State in Cairo, 1974-98*, School of Oriental and African Studies, University of London, 2007, p. 143.

nating from the negligence of the government, the other from the mechanism of allocating aid.

W.J. Dorman shows that although the government can build infrastructure and buildings, it is less adamant about accepting the idea of reforms by donors, thus the ease in creating projects "hard" and not "soft" as examined earlier. It is in this dimension that the "politics of neglect"[30] is most apparent. Rehabilitation projects, designed jointly by donors and local actors involved, imply greater citizen participation. However, the government sought rather to ensure his power with customers seeking favors and protection. Cooperation is limited by the state, which will not support the project that could affect the established order. The author then returns to the waves of developmental assistance in Egypt, made not too conditional exchange for the respect of treaties of peace and political stability in the region. Projects were often chosen for their ability to absorb large amounts rather than developing long-term growth. Aid had little leeway in strengthening governance in Egypt and the concept of development has mainly been expressed as "build this for us".[31] The government is not the only actor responsible, donors are as well: a large project can show visible signs of engagement in the country. However, they cannot buy political change. W.J. Dorman then denounces the infusion of aid, with which flowed a stream of experts, engineers and foreign consultants. They have enabled authorities to dispense with developing Egyptian professionals, which in turn could become a force to protest against the authoritarian exercise of power. It is important to note that, according to our interviews, this is becoming less true. Project ownership and a trend of replacing foreign personnel with local staff may continue.

B. A Progressive and Pragmatic Ownership within International Agencies

If you ask the cooperating actors in the field, the concept of sustainable development is broken down into multiple variables reflecting ownership of sustainable development on the ground, at for these players.

It is important to highlight that the replacement of international staff by local staff to form a local expert able to train other colleagues in turn. Two key notions are continually referred to: "capacity-building" and "ownership".

30 *Ibid.*

31 *Ibid.*

The sustainability of the project, its integration into the local context, and the involvement of all relevant actors have become key considerations:

> "When we talk about pilot projects in particular, we are concerned about Egyptian ownership [...] How do we talk about sustainability? We are trying to implement something applicable, and not high-tech systems from Europe and the United States that we could imitate here, no [...] We cannot copy solutions, finding applicable projects here is our main goal. This is what we make out of sustainability." Mohamed Fathy, sustainable transport project, UNDP (United Nations Development Program)

> "Sustainability depends on many counterparts that must be integrated with other relevant actors, from the beginning, this could define sustainability. [...] I saw recently, during the last 7-8 years of my career in development, positive change within civil society, taking into account changes in the representation of society, capacity building, and staff training." Khalil Shaat, GIZ (Gesellschaft für Internationale Zusammenarbeit)

In the teams working in urban cooperation, within donors and NGOs, sustainability is also reflected by a better integration of local staff as a direct result of "capacity building".

Sustainable development is a term used in practice for effective use and understanding of the field, and its word for word is mostly confined to official meetings and communication, essential to new projects. Implementation of sustainability requires a term more "practical" or "pragmatic", going through steps (assessment, ownership, impact studies, capacity building, sustainability...) that shape the sustainability.

There was also a desire on the part of donors to work together. Monthly meetings are held, organized by the Group of Partners for Development, creating working groups and facilitating communication between stakeholders. The European Union encourages the consolidation of donors: it has set up a financing instrument which forces players to join together in order to receive grants to reduce their lending rates. This new tool is transferred by the "neighborhood facilities", which can make the AFD, the European Investment Bank (EIB) and the German Development Bank (KFW) combine, creating a dynamic group having different expertise complementing each other.

C. The Usefulness of an Unavoidable Term?

There is therefore a development jargon that allows the use of key concepts for communication and projects' funding. This jargon has been simplified later and paved the way for a very pragmatic implementation. What does the systematic use of sustainable development in the cooperation involve? It appears as an unavoidable and useful term as it gathers different concerns – not necessarily new – under a single term.

If the discovery of "sustainable development" is not new in regards to the themes it conveys, it has the merit of providing a common concept combining local and global concerns, focusing on the need to link areas to achieve a more just and effective development. It allows you to manage a project with greater anticipation and regroups actors together around a common goal, without being too concise so that everyone finds his place. This is a language development. Speeches heard in meetings and those of field actors are therefore not at disconnect in regards to the language used on the field, both complementing each other.

It is key to point that if the use of the same term leads to a positive use of concepts, it is not entirely altruistic. This term is part of a rhetoric that has become an appliance which can give merit and legitimize a project to capture funds.

Development actors, through their interaction in the sphere of cooperation, their formation and origin, have appropriate themselves, according to their professional role, and the duration of sustainable development. Generic approaches such as the three-dimensional approach or preserving resources for the future, are indeed a minority ones compared to the specific approaches of sustainable development, delivered in a concept of project management. The adaptation is quite technical, relying on human capacity in general. There is not, in this circle of actors, a specific definition of sustainable development as such: one would expect that respondents mentioned the lack of water, climate, and desert environment. This might have been the case if we had scientists or researchers. Sustainable development is declined according to the scope of action.

If it is easy to address sustainability among professionals, the issue of adapting it to everyday life another matters.

V. In Search of a Sustainable City in Egypt

A. New Towns in Egypt, Quantitative Development

Nestled between a desert and the fertile lands of the Nile she tries to not consume the metropolis, suffering from various ailments such as pollution or congestion, looking for a way to grow without stifling. Grazing 20 million people now, projecting 35 to 38 million by 2050 if growth remains on the same pace, the Egyptian capital city presents many challenges commensurate with its size.

The various development plans in Cairo, starting in the 1970s, wanted to respond to the increasing population, by moving urban growth outside the city. It began with the creation of new towns far enough Cairo in order to develop a business hub without turning into a dormitory. Located about thirty to a hundred kilometers around Cairo, Ten of Ramadan, 6th of October and Sadat were created there over thirty years now. In the 1990s, the government developed "satellite cities", as Al-Qattamia, Chorouk Al-Sheikh Zayed, closer to the metropolis, to implement a social housing. The goal was quickly forgotten when the development was left in the hands of private developers. The use of the private sector is a consequence of liberalization and structural adjustment, which led to the withdrawal of the state. Broad land perimeters were indeed sold to promoters, following a commercial logic which guides the composition of the territory.

The results are not encouraging: the creation of dormitory cities dependent on the economic activity in Cairo, of "gated communities" reserved for the rich, quickly growing like around Cairo. The dense core of Cairo is now surrounded by buildings scattered without apparent unity, hardly connected to the city center and provides little opportunity for employment. Automobile dependence that characterizes these peripheries leads to more infrastructure dedicated to the car (overpasses, four channels), which continues to divide an already fragmented urban space.

Sustainability seems to be conspicuously absent from this development, despite the emergence some time ago of districts labeled as ecological neighborhoods. These neighborhoods, towns and cities require a high energy, fuel for transportation and water for the maintenance of green areas in the desert. With an occupancy rate well below of what was expected and a significant number of vacant housing, the results are far from positive. The latest generation of new towns, starting in the 2000s in Egypt, may also be confronted with the problems that have affected the outskirts of Cairo.

It is within this context and framework that the New Medina project takes place, focusing on the development of the city of New Fayoum, situated 15 km south-east of the Fayoum oasis itself, located about a hundred miles south of Cairo.

B. The New Medina Project

The New Medina project, launched in 2010 and co-financed by CIUDAD program,[32] focuses on the challenge of the development of

---

[32] The European Union launched in 2009 the CIUDAD program (cooperation in urban development and dialogue) to help local governments in Europe and in the European neighbourhood policy's partners countries to improve their capacity in urban sustainable development.

new towns in the southern shore of the Mediterranean Sea. As previously mentioned, this region is often characterized as political failures.

The main objective is to cooperate with Egyptian, Moroccan and Algerian new towns by providing an exchange of expertise strengthening the capacities of local teams. The projects brings three European partners together: the syndicate of the new municipality of Marne-la-Vallée Val Maubuée, Touiza Solidarité association and the European New Towns and Pilot Cities Platform (ENTP) as well as three partners from the southern shore: Al Omrane, operator for the city of Chrafate near Tanger, the town of Sidi Abdellah, close to Algiers, represented by the EPIC VNSA public company, and the Housing and Building national Research center (HBRC) which replaced the New Urban Community Authority after the Egyptian revolution.

Originally planned for a period of 3 years, the project has been allowed an additional year to complete its actions plan, since some activities have been delayed because of the Arab Spring.

Three panels of experts and several seminars have been organized on different themes ranging from governance to eco construction, energy efficiency of buildings, through new town's identity and its integration into the territory.

These successive and regular events contribute to the capitalization of knowledge on covered topics and in regards to the number of recommendations participants exchanged. New Medina's project ensures that the particular requests and wishes actually emanate from the Mediterranean cities, and it encourages, supported by the CIUDAD program, their financial independence on the project management.

This support is all the more favorable than the South/North financial flows represents, as for the people's mobility which is unfortunately constrained by difficulties such as the need to get a visa.

Partners do not always have the administrative and financial structures to first advance the necessary fund for the organization of events, and then obtain the reimbursement (absence of invoices, salary sheets, long and complex bureaucracy…).

It is not easy, for these countries, to adapt to the European system when they do not operate in the same way. The time spent on administrative difficulties is an obstacle for bodies which are not use to European projects. We can then wonder whether these projects are not too restricted to a limited number of beneficiaries. The support mechanism created by the CIUDAD program is one of the tools which have been created to solve these difficulties, representing a first step contributing to the sustainability of the projects.

The implementation of sustainable urban projects, as expressed previously through the New Medina's project, depends on the capability and the know-how of the different partners as well as on the institutional framework in which it is taking place. The integration of beneficiaries, this is to say the population, seems more complicated in Algeria due to a limited room for action of the planning authority.

In Egypt, the partner – being a national research center – is prone to this kind of reflection. Therefore, they could integrate teams of planners in its next workshop which will be held locally, following the last panel of experts held in New Fayoum in September 2012.

Course of the panel of experts in New Fayoum, September 2012

Pictures by the author

Conclusion

The emergence of sustainable development as a concept essential to cooperation projects is a "must-have" that meets both international injunctions and to the lost decades of development. Innovative instrument? Redesign of past actions? It is probably both. If the answer cannot be determined, it may be noted that it has created, available in "sustainable urban development", a space of confrontation and reflection. In fact, it tends to put the project in qualitative considerations rather than quantitative. If a concept is positive and unifying, the matter is not as simple. The notion of sustainability is evolving in an international context and in a fairly rigid financial and economic logic, does not necessarily combine well with the imperatives of sustainable development, and leaves little place to grow.

Understanding sustainable development in an alternative way than by its highly diagrammatic definition bringing it back to its three "pillars" shows that there was ownership. It is carried out by international agencies and institutions, but it is reused in a more pragmatic approach by development actors that have a direct connection with concrete action project. It builds itself mutually, and is justified by the position of: sustainability, financial sustainability, local ownership, capacity building, evaluation and impact studies, etc. This discourse is specific to project management. There is thus a generic term that is a staple of cooperation, and a series of practical terms on the field as a means to make it applicable in practice. Its definition, if it is not formulated as a commonly accepted form of development, more inclusive and participatory, will be found in words but declined in the construction of projects.

Thus, sustainable urban development makes its way to the southern shore of the Mediterranean, by adapting injunctions which are initially global, and eventually find a practical application in the field according to the priorities of the government.

Finally, even if one thinks sustainable development is only a little more than a recycling of old practices, it makes sense. Subway, bus, rehabilitation, while these are not new, they take a new and different meaning under prism of sustainability. This label of sustainable hides many other concepts taken in under one main concept. If there is a discrepancy between the way a project is mentioned and the way it is implemented, it means that all players have not yet appropriated the concept. The next bridge to cross to build sustainable urban development is in the hands of the authorities and civil society, both in the midst of transition following revolutions.

Bibliography

Allen, A., "Sustainable Cities or Sustainable Urbanization?", in *UCL's Journal of Sustainable Cities*, 2009.

Barbier, C. *et al.*, *Regards sur la Terre 2010: Villes, Changer de Trajectoire*, L'annuel du développement durable, Paris, Presses de Sciences Po, 2010.

Barthel, P.-A. and L. Zaki, *La Ville Durable au sud de la Méditerranée. Paroles de Chercheurs et de Professionnels*, Paris, Éditions de l'Aube, 2011.

Barthel, P.-A., "Premiers quartiers urbains "durables" dans les pays arabes: Enseignements et doutes sur une génération spontanée (Maroc, Tunisie et Égypte)", in Houpin S. (dir.), *Identification de critères de mise à l'examen de projets urbains durables dans le cadre de l'Union pour la Méditerranée*, rapport final du Plan Bleu, 2011.

Cavallier, G., "La nouvelle politique urbaine de la banque mondiale: un essai à transformer", in *Les annales de la recherche urbaine*, No. 86, Paris, 2000.

Chabbi, M., "La révolution, malgré la réhabilitation des quartiers populaires", in *Villes en développement*, bulletin du Partenariat Français pour la Ville et les Territoires, No. 91, June 2012.

Clerc, V., "Développement urbain durable et quartiers informels à Damas: évolution des paradigmes et contradictions des réformes", in Barthel, P.-A. and L. Zaki (dir.), *Expérimenter la "ville durable" au sud de la Méditerranée, Chercheurs et professionnels en dialogue*, Paris, Éditions de l'Aube, collection Villes et territoires, 2011.

De Ponte, G., "L'évolution du discours sur la ville: des institutions multilatérales d'aide au développement", in *Revue internationale des sciences sociales*, No. 172, 2002, p. 231-242.

Dorman, W.J., *The Politics of Neglect: the Egyptian State in Cairo, 1974-98*, Thesis, School of Oriental and African Studies, University of London, 2007.

Houpin, S., *Identification de critères de mise à l'examen de projets urbains durables dans le cadre de l'Union pour la Méditerranée*, Plan Bleu, 2011.

Hours, B., "Le développement durable, instrument d'intégration globale", in Martin, J.Y. (dir.), *Développement durable: doctrines, pratiques, évaluations*, Paris, IRD Éditions, 2002.

Mediterranean Institute: *Des politiques urbaines et territoriales encore fortement centralisées et pilotées par l'État*, accessed 26 July 2011, http://www.ins-med.org/2010/12/des-politiques-urbainesetterritoriales-encore-fortement-centralisees-et-pilotees-par-l%E2%80%99etat/.

Mediterranean Institute: *Les limites des outils de planification urbaine et territoriale*, accessed 26 July 2011 http://www.ins-med.org/2010/12/les-limites-des-outils-de-planification-urbaine-et-territoriale/.

Navez-Bouchanine, F., "Le développement urbain durable: "best practices" ou leurre méthodologique?", in *Espaces et Sociétés*, No. 131, 2007.

Project New Medina, compte rendu du troisième panel d'experts au New Fayoum, *Projet de la ville nouvelle de New Fayoum*, 2012.

Sims, D., "Printemps arabe et quartiers informels au Caire", in *Villes en développement*, bulletin du Partenariat Français pour la Ville et les Territoires, No. 91, June 2012.

Sims, D., *Understanding Cairo: The Logic of a City Out of Control*, AUC Press, 2011.

Tremblay, S., "Du concept de développement au concept de l'après-développement: trajectoire et repères théoriques", in Collection *Travaux et études en développement régional*, Université du Québec à Chicoutimi, December 1999.

United Nations, *Agenda 21*, World Summit, Rio de Janeiro, 1992.

Making Egyptian Cities Sustainable

Dream & Reality

Dr. Arch. Hend FAROUH

Associate Professor of Sustainable Architecture & Urban Development at Housing & Building National Research Center (HBRC), Egypt

Abstract

Though forty years of development efforts in desert areas have passed, Egypt is still lacking a successful strategy and methodology to establish new settlements in the desert. Such a strategy would need to aim at solving the challenge of increasing population, taking into consideration that about 96% of Egypt is covered by the Sahara desert, with the remaining, more fertile land concentrated along the Nile River. These issues have prompted the Egyptian authorities to develop plans and adopt policies to expand into the desert and creating new foundations for urbanization outside the inhabited area.

Twenty-two new cities have been established in Egypt from 1977 till 2000 which are divided in terms of locations and functions into: satellite, twin and independent cities. These cities are facing economical, social and environmental challenges. Moreover, most of them have not yet achieved their main goals due to the lack of comprehensive integrated development approach.

Egypt is in a real need for a paradigm shift in urban planning and designing of its new cities in order to achieve sustainable development, breaking away from the conventional patterns of urban development.

There is not agreement worldwide on a definition for what a sustainable city should be or what components should be included. On the other hand, this is still a new concept in Egypt and has no clear definition or settled criteria till now.

This study will present the challenges facing Egyptian cities and will explore the current situation in the new cities besides, the reasons behind not realizing their targets in order to propose an approach for sustainable cities in Egypt. This approach is dedicated for the existing new cities of the first and second generations, what is more, it will be for the intended planned new cities of the third generation as well as the new settlements that should be created in the desert area; taking into consideration the different dimensions of sustainability and their specific manifestations on real ground.

I. Challenges Facing Egyptian Cities

Egypt is facing a great challenge in the combination of continued rapid population growth with severely constrained fresh water and energy resources. The overall population of Egypt increased from 22 million in 1950 to 83 million today, and is likely to increase to above 120 million by 2050. Nevertheless, a great majority of the Egyptian population concentrate along the narrow areas of the Nile Valley and Delta. Approximately 95% of the population uses only about 5-7% of the total land area. The World Bank report in 2008 stated:

> Egypt is facing a daunting urban challenge. In the next fifteen years, Egypt's population is expected to increase by 27 million inhabitants to reach over 100 million. Most of this increase will occur within urban areas and in the "urban villages" within urban agglomerations. Accommodating this huge population increase in such a short period of time is a major challenge for the Government. Urban economies will need to generate a large share of the approximately 700,000 jobs needed for new entrants to the labor market each year, especially for limited income groups. New urban dwellers will also need access to affordable housing, and the GOE will need to deliver related urban infrastructure and public services. It is estimated that 300,000-400,000 housing units are needed annually for the coming 15 years, two-thirds of which are for limited income households. In addition, a majority of Egypt's existing urban population is of limited income and suffers from a lack of appropriate urban services, high levels of unemployment, and inadequate and crowded housing. Over 16 million urban inhabitants live today in informal and squatter settlements.[1]

The urban population in Egypt is unevenly distributed among the 219 cities, in which those inhabitants live; Egypt's two primary cities Cairo and Alexandria comprise 43% of the total urban population (17% of Egypt's overall population) while 77 cities comprise 4% of the urban

1 The World Bank, "Arab Republic of Egypt, Urban Sector Update", Cairo, April 2008.

population.[2] Regional disparities and urban primacy are associated with social ills and economic problems, such as unemployment, poverty, environmental degradation, and denying marginalized sub-populations access to power and wealth. It is also often equated with poor use of national resources, ruralisation of urban areas, and excessive growth rates of villages without provision of adequate infrastructures and social services.[3]

Youth unemployment is one of the reasons behind the 25th January's revolution and it increases. The unemployment rate in Egypt has been last estimated by 12.6% in the second quarter of 2012. Unemployment is one of the main challenges that face the Egyptian Government today. Hence, it should be taken into consideration in the urban planning of the new cities.

In addition, Egypt is facing a crucial energy problem. Natural gas consumption was nearly doubled over the last decades and reached 1.6 trillion cubic feet in 2010. Total petroleum consumption has risen by about one-third over the same time period.[4]

Similarly, electricity generating capacity has grown steadily to keep up with the peak demand, almost doubled between 1990 and 2009 from 12,230 MW to 23,502 MW. Table (1) shows that 47% is for residential buildings electricity consumption, then 20% for industrial electricity consumption and 12% for Governmental buildings electricity consumption.[5]

Table 1. Electricity consumption by sector, 2010

Sector	Electricity Consumption (%)
Residential	47
Commercial	3
Industrial	20
Governmental	12
Public Lighting	9
Agriculture	4
Other	5

2 General Organization for Physical Planning, Development and Construction Map for Egypt 2017, June 1998.

3 Goell, E. *et al.*, *Sustainable Cities in Egypt. Learning from Experience. Potentials and Preconditions for New Cities in Desert Areas*, The Egyptian Cabinet, Information and Decision Support Center, Center for Future Studies, September 2009.

4 US Energy Information Association, http://www.eia.doe.gov.

5 African Development Bank, "Arab Republic of Egypt: Power Sector in Brief", 2010.

According to the Shura Council's Financial and Economic Affairs Committee report, the subsidies allocated to petroleum products have increased from nearly LE 290 million in 1970 to more than LE 40 billion at 2011. The government has been pushing to reduce the energy's subsidy. Egypt is relying mainly in its consumption on Oil & Natural gas although some estimates indicate that indigenous natural gas and oil reserves, on which Egypt's electricity generation currently relies, will run out in 30 or 40 years. Thus, the transition to alternative energy sources becomes a pressing need to avoid stagnant economic development.

The present energy strategy (the resolution adopted by the Supreme Council on energy in 2007) aims at increasing the share of renewable energy to 20% of the energy mix by 2020. This target is expected to be met largely by scaling-up of wind power; as solar is still costly and the hydro potential is largely utilized. The share of wind power is expected to reach 12%, while the remaining 8% would come from hydro and solar. This translates into a wind power capacity of about 7,200 MW by 2020. The solar component is at this stage considered to start with 100 MW of CSP and 1 MW of PV power.

Besides the previous challenges, Egyptian urbanization creates some environmental challenges as stated in a report by the UNICEF in 2010; which was addressed "Egypt's progress towards achieving the Millennium Development Goals with regards to (Ensure Environmental sustainability)."

Due to rapid urbanization, industrialization and heavy population density along the limited and confined green valley of the River Nile, several environmental problems emerged, affecting both urban and rural sectors at the macro as well as the micro levels. The lack of environmental awareness aggravates the problem of natural resources degradation and generates sanitation, hygiene and environmental problems at the community level.

In spite of high stated potable water and sanitation coverage along with the high values required by the Egyptian codes in that sense, in reality the leakage and wastage of provided water is excessive and the final disposal is not environmentally friendly.

Moreover, the waste management system of both hazardous and non-hazardous municipal, agricultural, construction/demolition and industrial waste is inefficient in Egyptian cities. An average of 15.3 million tons of municipal solid waste is generated each year, out of which almost 2.5 million tons remain uncollected and lack appropriate sanitary landfills for their final disposal.

Different development projects in Egypt have dealt with these challenges. The following table summarizes some of these development projects (1986-2012) showing the success and the outcome of each project:

Table 2. Summary of some development projects in Egypt (1986-2012)

Name of the Project	Abbreviation	Duration	Funding org.	Success	Nature & outcome
Basic Services for Villages	BVS	1978-1986	USAID	Field success	Service delivery oriented
Local Development	LD	1986-1994	USAID	Good	Services-limited participation
Supporting Democracy	DEM	Have Not Yet Been Implemented	USAID	Have Not Yet Been Implemented	Political issues
Integrated Rural communities	Shorouq	1992-2002	Egypt+	Good	Aborted
Total Environmental Care	TEC	1998-2001	UNICEF	Good	Aborted
Local Development Fund	LDF	1987-now	Egypt+	Good	In Progress
Social Development Fund	SFD	1992-now	Donors	Weak success	Lack of Egyptian vision for development
Egypt's Project	MISR	2003-2006	UNDP	Failed	Aborted and cancelled
Thousand Villages Project	1000V	2006-now	GOE	Stopped	NA
Urban Planning for Villages	VUPP	2004-now	MOH	Useless	NA
New Cities	NC	1978-now	MOH	Failed	Objectives unaccomplished
Million Population Cities	NMC	2008-now	MOH	NA	NA

Source: R. Haggag, A Presentation for Youth Development Project "PLEV", HBRC, 2011

II. Egyptian Cities: an Incomplete Story!

Starting in the mid-1960s, vast unrevised and underserviced informal settlements proliferated illegally on agricultural lands on the urban fringes and on state-owned desert lands at the edge of the valley. Between 1976 and 1994, the country lost an average of 8,000 ha of agricultural land yearly towards urbanization into the desert. It has recently turned to land-based financing to cover the high cost of extending services to the desert to attract private developers to invest in the development of desert sites.[6]

The authority of the new urban communities, NUCA, founded in 1979, is the developer of the new cities in Egypt. The 22 new Egyptian cities created between 1977 and 2000, which differ via their location and function, and could be grouped into three categories: Satellite towns (built around Cairo); twin cities, which are mostly extensions of existing towns; and "independent" cities, which are distant enough from other urban entities to develop their own economic base.

Proliferation of informal settlements on agricultural lands

© EEAA, 2010

The World Bank report in 2008 showed a skeptical assessment of the idea of developing and populating desert areas on a big scale stating:

> Migration to the new desert urban communities has been practically insignificant. For example, the total population of all the new towns and settlements in Cairo's desert in 1996 did not exceed 150,000 persons, one part of this population was in the 15th May city; a public housing project which was grafted onto the Helwan suburb. For comparison, over the 1986-1996

6 UN Habitat, "The State of the Arab Cities 2012: Challenges of Urban Transition", 2012.

periods the population of Greater Cairo grew by over 2.1 million persons. In other words, by 1996 all the new towns and settlements around Cairo did not absorb the equivalent of 6 months of Cairo's growth! In 2006 the Census recorded only 602,000 people living in the new towns around Cairo, attracting 451,000 persons or only 13.8% of the 3 million people added to all Greater Cairo over 10 years. At the national level, the new desert communities have even had a less demographic impact. In 2006 the population of all Egypt's new towns (did not exceed 766,000 persons, or only 1.06% of Egypt's total population. And over the 1996-2006, period all new towns only absorbed 4.3% of the nation's population increase.[7]

Although the objective pronounced in all studies of new cities indicates that the construction purpose is to mitigate the population pressure of the existing cities (especially Cairo and Alexandria), the housing problem is growing more complex. The number of Egyptian households is rising by around 100,000 per year without a corresponding increase of housing units. On the other hand, the existing cities are still suffering from unemployment. It was further proved that most of the industrial projects established in such cities failed to make a tangible change in the labor market mechanism. Moreover, most of the working manpower is recruited from other places (not residents).

Studies indicate that the percentage of population attraction into the housing units in the new cities failed to fulfill the target. For instance, 40% of the housing units remained un-occupied in these 22 new cities due to several reasons, *inter alia*: rejecting governmental housing for increased amount of rent or ownership installments that are incompatible with the income. Moreover, 25% of commercial areas are not used because commercial and public services cannot be performed efficiently.

The housing problem in Egypt is far more complex than simply an insufficient number of available units. The World Bank affirms that there are 2.65 million unused housing units, nearly 28% of the total number of urban housing units.[8]

In addition, it is estimated that at least 45-50% of the urban housing stock in Egypt faces problems such as vacancy, rent control, and informality. Informal housing stock represents great potential for providing affordable and flexible alternatives for low and middle-income groups, especially among youth and newly established families, since it offers a wide variety of financial models starting from an upfront down payment.[9]

---

7 The World Bank, "Arab Republic of Egypt, Urban Sector Update", Cairo, April 2008, p. 12.

8 The World Bank, "Analysis of Housing Supply Mechanisms", October 2006.

9 Egypt Human Development Report, 2010.

The public and private real estate markets operating in desert towns have failed to create affordable, appropriate and sustainable housing. There are several explanations for this. Let us first consider government public-housing programs. In these programs, housing has none of the advantages of other areas that arise from high density, diversity, access to informal jobs and markets, affordability and proximity to family and friends.[10] Housing research since the 1980s in Egypt has focused on the problems of informal areas, despite that these areas grow faster than any other housing sector in the country.

In 2009, Edgar Goell *et al.*, in a report *Sustainable Cities in Egypt, Learning from Experience: Potentials and Preconditions for New Cities in Desert Areas*, surmised the reasons of the slowdown of human settlement in the Egyptian new cities and un-accomplishment of its desired goal in the followings:[11]

– Fragmented urban growth: It was clear that the growth of housing areas and service centers was fragmented. In this context, a large number of incomplete public and private housing projects in many neighborhoods might be found, besides the unplanned distribution of service facilities on the three levels, namely: neighborhood, district and city.
– Lack of funding: Achieving economic development was the main goal of constructing new cities to allow an expansion of industrial and agricultural sectors aiming to create job opportunities and an increase in both production and national income. It was necessary to provide funding sources to cover the projects of infrastructure, social services and housing. However, the situation can be explained as such, an increased number of spacious cities and lack of self-funding.
– Shortcoming of implementation: Implementation problems of the infrastructure, housing, industrial and services facilities can be summed up in poor construction capacities affecting the programs of development for new cities.
– Shortcoming of management: The overall deficient picture of new urban cities and communities in Egypt clearly clarifies, in addition to poor financial resources and building capacities, a shortcoming of administrative processes in the relevant bodies. This problem negatively affected the optimal use of available resources and capacities even if they were limited.

---

10 Tarbush, N., "Cairo 2050: Urban Dream or Modernist Delusion?", in *Journal of International Affairs*, 65:2 (Spring/Summer 2012), pp. 171-188.

11 Goell, E. *et al.*, *Sustainable Cities in Egypt. Learning from Experience. Potentials and Preconditions for New Cities in Desert Areas*, The Egyptian Cabinet, Information and Decision Support Center, Center for Future Studies, September 2009.

- Lack of integrated development approach: The problems of the Egyptian experience of desert development, in its early phases, stemmed from the lack of comprehensive development approach. In this context, the development process adopted a unilateral dimension.
- Random selection of locations: This problem can be clearly spotted because determinants, capacities, environmental and construction opportunities, and overall features of the sites were not clear.

The main reasons for these failures could be concluded as: lack of vision and holistic approach in the development of these new cities; interference of political decisions especially in site selection of the new cities which can lead to random selection of locations; ignoring public participation in the planning process where the urban design does not pursue the new inhabitants' needs; culture and life style; lack of Job opportunities and public services; lack of urban marketing, etc.

III. What Is a Sustainable City?

There remains no absolute agreement upon a definition for what a sustainable city is, or complete agreement upon a paradigm for what components should be included.

Generally, developmental experts agree that a sustainable city should meet the needs of the present without sacrificing the ability of future generations to meet their own needs. The ambiguity within this idea leads to a great deal of variation in terms of how cities carry out their attempts to become sustainable.[12]

When the concept of sustainability extended from the global to the local level, the term "sustainable cities" becomes another buzzword among academics, politicians, media and general public. Although sustainability is initially a global concept, an important scale of study is at the local level about urban sustainability.[13]

In their 2009 book "*A Renewable World: Energy, Ecology, Equality*" Herbert Girardet & Miguel Mendonca present that cities should be a dynamic and ever-evolving eco-technical systems which can help us formulate strategies for a sustainable urban future. For their long-term viability we would be well advised to model our cities on the functioning of nature's own highly complex ecosystems, such as forests or coral reefs, in which every output discharged by an organism becomes as input that renews and sustain the continuity of the whole. Developing

---

12 Magilavy, B., "Sustainability Plan". Sustainable City, http://www.sustainable-cities.org, retrieved 6 December 2011.

13 Eden, S., "Environmental Issues: Sustainable Progress?", in *Progress in Human Geography*, 24:1, 2000, pp. 111-118.

environmentally sustainable cities is one of humanity's greatest challenges for the new millennium. Across the world, a revolution in 'future proofing' our urban systems has started, assuring high level of energy efficiency and rapidly switching to renewable energy technology. The sustainable city principles are:[14]

- Compact urban development
- Renewable as primary energy supply
- Small ecological footprint
- "Circular urban metabolism closed cycle"
- Biodiversity in landscape design
- A city embedded in farmland.

IV. Towards Egyptian Sustainable Cities

Developing world cities face different sustainability challenges to cities in the Developed World. Each country has its own challenges beside the international challenges that should be reflected in the approach and implementation of sustainable cities.

Herbert Girardet & Miguel Mendonca, 2009 mentioned:

> It is what was said by urban Analysts that Cities Should be seen as the places where solutions to the world's environmental and climate problems can most moonrise ask Implemented Because as places where most people live competitive together They have the potential to make efficient use of resources. Recycling materials, conserving rather than burning fossil fuels and tapping into renewable energy can create many local jobs. In many cities, large numbers of people cannot find jobs because economic globalization has caused relocation of manufacturing jobs to other parts of the world. We need creativity and initiative at the local level, but we also need appropriate national policies to enable useful things to happen locally. Without enlightened national policies in each country, the changes will not happen for incidental vast enough.

The sustainable cities approach should start from the local to the global level. Nowadays, the concept of sustainable development is not implemented in Egypt's city planning despite the presence of a ministerial declaration on sustainable development, issued in Cairo on the 25th of October 2001. On the regional level, the League of Arab States adopted a comprehensive regional approach, through the Council of Arab Ministers Responsible for the Environment and other specialized ministerial councils and in cooperation with international, regional and

---

14 Girardet, H. and M. Mendonca, *A Renewable World. Energy, Ecology, Equality. A report for the World Future Council*, Green Books ltd, UK, 2009.

Arab organizations which aims at developing a regional program for sustainable development. Moreover, a sustainable development initiative in the Arab Region is being carried out aiming at addressing the challenges facing the Arab countries to achieve sustainable development. It asserts the commitment of the Arab countries to implement Agenda 21 and the included development objectives in the Millennium Declaration and the outcome of the World Summit on sustainable development, taking into consideration the principle of common but differentiated responsibility.

Managing the Consequences of a largely unplanned transition and decades of poor governance, preceding democratization in Egypt, connected a shift towards sustainability using a holistic approach to achieve integrated sustainable development.

Tackling the sustainability of Egyptian cities has two levels:

- Dealing with the existing cities, including the 22 new cities which represent as mentioned previously about 5-7% of total Egypt's area.
- Dealing with the new communities that should be established in the rest of the desert area which represents 93-95% approximately.

A. Dealing with Existing Egyptian Cities

This study focuses on the new existing 22 cities which have almost the same challenges and problems that were presented previously. Dealing with these cities from the sustainable point of view should have a list of priorities in three main levels, taking into consideration: the socio-cultural aspects, involving society and all stakeholders as well in all the processes of reforming these new cities. These three main levels are as follows:

On the economical level:

- Create job opportunities by reconsidering the economic rationale behind each new city and the available local resources and how it could be reformulated to establish a strong economic base in each city.
- Enhance the capacity development and vocational training for the residence and the new comers of these cities.
- Creating public awareness of the economic benefits of sustainable urban development.
- Assuring effective knowledge & innovation exchange within and between cities.

On the environmental urban planning level:

- Formulating strategies for better linkages and integration with existing nearby urban agglomerations. New cities cannot be treated as isolated geographic entities.

- Addressing public transport problems, with solutions tailored to each city and its geographical context.
- Enhancing the primary and secondary services for each city which is one of the major problems that affect attracting people to new cities.
- Enabling renewable energy and energy efficiency development through developing new approaches in urban governance and planning.
- Focusing on the new approaches for water efficiency and waste management for each city according to the existing situation as they represent the main challenges.
- Raising public awareness of the cities role in solving the problems of water, energy, waste management and food.

  On the governance & administrative level:[15]

- Creating a marketing body in each city that comprise an information center covering the housing, populace, investment situations and all other related basic data. This body develops an awareness media plan – on both the local and regional levels – regarding the benefits, investment and work opportunities. The plan must cover potential incentives, exemptions and services introduced in the given city.
- Setting a more liberalized administrative system and enabling the cities an increased level of independence. This will achieve the following:
  - The local authority of the city will be in full control of planning and implementing civil life requirements.
  - The community of the city will be empowered for an effective participation in constituting the principles of democracy.
  - It is expected to accomplish an increased level of decentralization. As such, the relation between the city and higher administrative levels will be restricted to regional and national affairs requiring State sovereignty.

Each point of the previously discussed should be addressed in details for each city through involving all stakeholders according to the challenges, current situation, local resources, potentials and the desired outcome.

B. Dealing with New Settlements

Egypt already suffers from slums. The most significant shift to date is still the development of suburbs and creating new communities in the desert area to reduce unplanned urban sprawl. It is essential to develop a

[15] Goell, E. *et al.*, *op. cit.*

holistic and comprehensive understanding of the Egyptian desert nature and the socioeconomic systems before getting into constructing new urban cities and communities in the desert. In this regard, mechanisms of new concepts of sustainable development and green construction would need a motivating State.

Each new community should start by creating a vision that is embraced and actively promoted by all of the stakeholders and key sectors of society and government. On the other hand, urban planners and architects should call the vernacular urbanism concepts that has arisen from and which are expressions of the Egyptian local culture and ethos. Every new city can re-imagine its future by first connecting with the inherent vernacular urbanism of its past. The main features of the vernacular cities in Egypt were compact urban planning, narrow irregular street pattern, courtyard houses and courtyard forms and interior spaces that look like local market (souk), medieval walls and gates, covered markets that was convenient to the climatic conditions and that met the needs of residents.

In addition, sustainable city development projects should put emphasis on city marketing, awareness raising and capacity building in the early stages of project planning, design and implementation taking into consideration the main issues that could attract people to leave the Nile Valley and move to live in the desert which is the corner stone of creating new communities in the desert.

Egypt has started recently to move forward towards the concept of sustainable cities. The Housing & Building National Research Center in Egypt has lead a national project to establish a rating system for sustainable community called GPRS "Green Pyramid Rating System for Sustainable Communities" which includes seven categories: sustainable site & ecology, sustainable urban planning & design, water efficiency, energy efficiency, material & resources, management and innovation & added Value. Each category provides definitive criteria and benchmark for good practice that enables new communities in Egypt to be assessed for their green credentials through a credible, challenging and transparent national environmental rating system.

Many models were introduced for the sustainable cities. A model of "Four E's" of sustainable cities in Egypt as shown in Figure (1) was developed incorporating elements of the "Three E's" model of sustainable city by the Regional Workbench Consortium (RWBC) which is a collaborative network of university, industry, government, and community organizations dedicated to enabling sustainable city-region development. This model of "Four E's" is introducing a wider concept for sustainable cities in Egypt which consists of (Eco-urban systems, Economy, Equal life quality and Establishment).

Figure 1. Four E's of Sustainable City Planning in Egypt

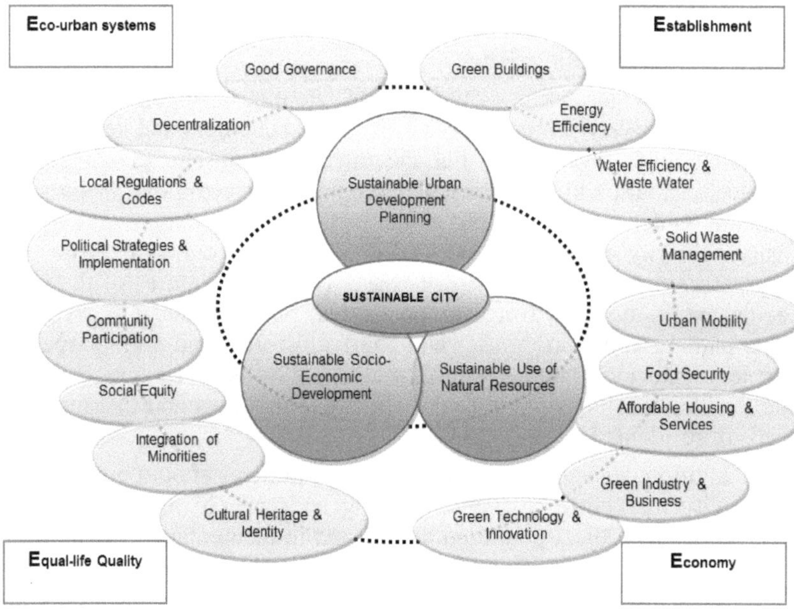

One of the main obstacles that faces the concept of sustainable cities in Egypt, is the existing term of reference (TOR) for planning or re-planning the Egyptian cities which deals typically with all the Egyptian cities in different locations regardless of different climatic zones, local resources, and culture, etc. The TOR set out a road map and determines a detailed collection of the contractual conditions that presents for the required jobs. Moreover, it shows how the scope will be defined, developed, and verified, besides developing a common understanding of the scope among stakeholders. In order to meet the required criteria, success factors/risks and restraints should be the fundamental keys.

Creating a detailed term of reference for a sustainable city is a critical issue, as it defines the vision, objectives, scope and deliverables, stakeholders, roles and responsibilities, resource, financial and quality plans and work breakdown structure and schedule.

Egypt is in a real need to develop a TOR that satisfies the requirements of the Unified Building Law No. 19, 2008, chapter (1), rule (2) which states:

- Sustainable urban development is a management of the physical development process through optimal utilization of available natural resources for current generation without affecting opportunities for future generations

– General strategic plan for the city and the village should identify the future needs for economic, social, environmental, physical development plans and projects to reach sustainable development at the local level within the vision of the governorate level. It should also indicate the land use budget, building conditions, implementation programs, priorities, mechanisms, and financing sources.

Table 3. Proposed categories and sub-categories for New Term of Reference (TOR) of Sustainable City Planning in Egypt

Categories	Prerequisite	Economical Development	Social Development	Environmental Development	Urban Planning	Management
Sub-Categories	Vision of development	Economical Base	Demographic data	Energy	Urban structure	Planning regulations
	Regional plans	Economy Growth	Basic services	Water	Services & Facilities	Construction regulations
	General strategic plan	Consumption/ production pattern	Social services	Air	Economic activities distribution	Implementation programs
		Food security	Cultural Heritage	Soil	Zoning	Implementation priorities
		Housing provision		Climatic conditions	Infrastructure	Implementation mechanisms
				Biodiversity		Sources of funding

[white box] Elements exist in TORs (General Organization for Physical Planning-GOPP and Habitat)

[grey box] Suggested elements according to the Egyptian Laws and the International Signed Treaties (Law 119 of 2008, MDG, Agenda 21, CRC, HRW)

R. Haggag, A. Galal, S. Eldefrawi, N. Osama, R. Adham, New Medina Project, 2012

Table 3 shows a proposal of categories and sub-categories for new TOR of sustainable city planning in Egypt which was developed in the frame of the "New Medina Project: From pilot towns to sustainable towns: reinventing new towns".[16] These categories and sub-categories

[16] New Medina Project, http://newmedina.eu/.

were proposed according to a comparative analysis of the two existing TORs (General Organization for Physical Planning-GOPP & UN Habitat), the requirements of Unified Building Law (119/2008), the signed treaties by Egypt and basically the MDG Eight Goals, HRW, CRC additionally Agenda 21.

The planning and design of new settlements in Egypt should take into consideration the proposed categories (economic, social and environmental development, urban planning and management) and its subcategories.

Conclusions

Egypt is facing major challenges; a combination of continued rapid population growth, political transition, high unemployment rate, poverty, severely constrained fresh water and energy resources, inefficient waste management and air pollution. Hence, the need of a paradigm shift in urban planning and design of its new cities is of utmost necessity. This must take into consideration that the conventional patterns of planning have not achieved the main goals of attracting Egyptians to new cities, especially in remote desert areas due to the lack of a comprehensive integrated development approach. The concept of sustainable cities in Egypt is still new and it should be reflected in the term of reference of planning new cities and in re-planning existing cities.

Dealing with existing new cities, urban planning should have a list of economic, environmental and social priorities according to the current situation, local resources and challenges. It should take into consideration the socio-cultural aspects and the importance of involving all the stakeholders and inhabitants in the development process.

On the other hand, planning and design of new cities should depend on a holistic and comprehensive understanding of the Egyptian desert nature and the socioeconomic Egyptian systems. Each new community should start by creating a vision that is embraced and actively promoted by all stakeholders and key sectors of society and the government which should recall the lessons of vernacular cities to be connected with the future. In addition, this should be reflected into the different dimensions and categories of sustainable cities.

Abbreviations

GOE	Government of Egypt
HBRC	Housing & Building National Research Center

Bibliography

African Development Bank, "Arab Republic of Egypt: Power Sector in Brief", 2010.

Eden, S., "Environmental Issues: Sustainable Progress?", in *Progress in Human Geography*, 24:1, 2000, pp. 111-118.

Egypt Human Development Report, 2010.

General Organization for Physical Planning, Development and Construction Map for Egypt 2017, June 1998.

Girardet, H. and M. Mendonca, *A Renewable World: Energy, Ecology, Equality, A report for the World Future Council*, Totnes, Green Books ltd, 2009.

Goell, E. *et al.*, "Sustainable cities in Egypt, Learning from Experience. Potentials and Preconditions for New Cities in Desert Areas", The Egyptian Cabinet Information and Decision Support Center, Center for Future Studies, September 2009.

Magilavy, B., "Sustainability Plan", Sustainable City. http://www.sustainable-cities.org, retrieved 6 December 2011.

New Medina Project, http://www.newmedina.eu/.

Tarbush, N., "Cairo 2050: Urban Dream or Modernist Delusion?", in *Journal of International Affairs*, 65:2, Spring/Summer 2012, pp. 171-188.

The World Bank, "Analysis of Housing Supply Mechanisms", October 2006.

The World Bank, "Arab Republic of Egypt, Urban Sector Update", Cairo, April 2008.

UN Habitat, "The State of the Arab Cities 2012: Challenges of Urban Transition", 2012.

US Energy Information Association, http://www.eia.doe.gov.

Authors

Pascaline Gaborit

Pascaline Gaborit has 10 years of experience working with European local authorities, as Director of the ENTP: European New Towns and Pilot Cities Platform (a network of European local authorities). She coordinates exchanges, events and expert panels, as well as giving university lectures. She also manages EU co-financed projects and conducts research on urban development (particularly regarding New Towns), and on the different policies, actions and roles of local authorities (particularly regarding planning, management, environment, economy, communities and culture). As such she is convinced that cooperation and exchange between local authorities is very important.

Recent projects include cooperation with Chinese and Indian New Towns (EAST project: http://www.east-project.org), with Southern Mediterranean New Towns (http://www.newmedina.eu), projects on sustainable renovation (http://www.greenov.net), Culture and development (http://www.award-project.eu) and migrants' inclusion in Europe (Labour plus).

Pascaline completed a PhD in Political Science, speaks several languages and has participated as an expert/consultant for various European and international programs and reports. She is the author of various books, articles and conference papers including *European New Towns: Image, Identity and Future Perspectives*, published by Peter Lang, 2010.

Rachid Sidi Boumedine

Rachid Sidi Boumedine currently works as an expert and associate director for research at CREAD (Centre for Research in Applied Economics for Development) in Algiers. He has previously taught Sociology (Methodology and Urban Sociology) and conducted research at the University of Algiers, as well as carrying out work for CNRS and the IRD (France). He has directed various public research organizations and organizations for research on urban planning, territorial management and environment, as well as conducting numerous studies.

Deni Ruggeri

Dr. Deni Ruggeri is Assistant Professor of Landscape Architecture at the University of Oregon, where he is Director of the PhD Program. He is trained as an architect, landscape architect and urban designer, holding a PhD in Landscape Architecture and Environmental Planning from the University of California, Berkeley.

Dr. Ruggeri's research focuses on the interface between physical environment and human behaviour. His research investigates the influence that urban design and landscape architecture have on people's place identity and attachment to everyday neighbourhood landscapes in new communities. He was the co-editor of the Spring 2007 issue of "Places" Journal as well as being the author of various articles published in books, academic journals and conference proceedings. His most recent projects include the master planning, design and construction of the Jeffrey Open Space Trail in Irvine, CA, the residential communities of Ridgegate in Colorado, the Spanish Walk in Palm Desert, CA and a New Town center for Westport, California. Dr. Ruggeri is currently researching garden cities in various parts of the world.

Christian Horn

Christian Horn is an architect and urban planner. He obtained this double qualification in architecture and urban planning at the Technical University of Aachen in 1998. His agency, which was created in 2002, works in collaboration with other professionals for international clients both public and private, on projects ranging from the metropolitan to the architectural. He regularly publishes texts, essays and articles related to architecture and urban planning. Since 2005 he has been involved in the Ateliers Internationaux de Maîtrise d'oeuvre Urbaine (International Workshops for Urban Planning and Development), where he has become a member of the Committee for Scientific Direction and pilot in a workshop in Vitoria, Brazil, in 2009 and in La Défense, France, in 2011. Since 2007 he has taught at the School of Architecture in Paris-la Villette in the master in Architecture, Environment and Sustainable Development. He speaks regularly as expert in the competition EcoQartiers of MEDDTL, in the commission EUROPAN 11 and in other competition jurys.

Pierre-Arnaud Barthel

Pierre-Arnaud Barthel is a lecturer in urban planning at the University Paris-Est Marne-la-Vallée and LATTS. His recent publications concern the guidelines of the sustainable city in the Arab World: Barthel P.-A. & Monqid S., *Le Caire, réinventer la ville*, Autrement, collection Villes en mouvement, 2011; Barthel P.-A. & Zaki L. (dir.), *Expérimenter la "ville durable" au sud de la Méditerranée*, Éditions de L'Aube, 2011.

Lauriane Lahery

Lauriane Lahery joined the European New Towns and Pilot Cities Platform in October 2011 and is now working as a full-time project officer on various European projects. She completed a Master program in Geography with a specialization in "Globalization, spatial patterns and sustainable development in the southern countries" from Paris IV Sorbonne University, in 2010-2011. She wrote her Master thesis about sustainable development and international cooperation in Cairo under the supervision of Pierre-Arnaud Barthel, former head of the Urban Studies Department in the CEDEJ, a multidisciplinary research institution dedicated to the Arab-Muslim world studies.

Hend Farouh

Hend Farouh is a PhD architect and senior researcher at the Housing & Building National Research Center (HBRC) in Egypt. She has a postdoctorate in Green Technology for Sustainable Urban Planning and Design from Politecnico di Torino, Italy and holds PhD & M.Sc. in Sustainable Built Environment and Urban Development, Cairo University.

Dr. Farouh has more than 12 years of professional experience in the field of sustainable design and urban planning. She is specialized in transdisciplinary research in national and international projects and serves on several committees such as the "Bioclimatic Architecture in the Mediterranean Manual", "Green Pyramid Rating System for Sustainable Communities" and the "Egyptian Energy Efficiency Building Code". In addition, Dr. Farouh authored and co-authored many peer-reviewed scientific papers and has contributed to diverse conferences and workshops all over the world. In 2004 she was granted the "Honorary Reward" for the best achievement in Research in "Architecture, Energy & Environment" from Lund University, Sweden.